National Aeronautics and Space Administration

D0282430

Headquarters
Washington, DC 20546-0001

October 2008

Dear Reader:

Please note that this is an updated version of the 2007 edition of *William H. Pickering:
America's Deep Space Pioneer*, with typographical errors corrected. The content remains the
same. Thank you very much.

Sincerely,

Stephen Garber
NASA History Division

National Aeronautics and Space Administration

Headquarters
Washington, DC 20546-0001

October 2008

Dear Reader:

Please note that this is an updated version of the 2007 edition of *William H. Pickering: America's Deep Space Pioneer*, with typographical errors corrected. The content remains the same. Thank you very much.

Sincerely,

Stephen Garber
NASA History Division

William H. Pickering

America's Deep Space Pioneer

ISBN 978-0-16-081536-2

ISBN 978-0-16-081536-2

William H. Pickering
America's Deep Space Pioneer

Douglas J. Mudgway

The NASA History Series

National Aeronautics and Space Administration
NASA History Division
Office of External Relations
Washington, DC
2008
NASA SP-2008-4113

Library of Congress Cataloging-in-Publication Data

Mudgway, Douglas J., 1923-
    William H. Pickering: America's deep space pioneer/by Douglas J.
    Mudgway.
        p. cm. — (NASA history series ; Special Publication 2008-4113)
    Includes bibliographical references and index.
    1. Pickering, William H. (William Hayward), b. 1910-2004.
    2. Astronautics—United States—Biography. I. Title.
    TL789.85.P54M84 2007
    629.4092—dc22
    [B]
                                                            2007019158

Please note that this is a corrected version of NASA SP-2007-4113.

"To set foot on the soil of the asteroids, to lift by hand a rock from the Moon, to observe Mars from a distance of several tens of kilometers, to land on its satellite or even on its surface, what can be more fantastic? From the moment of using rocket devices a new great era will begin in astronomy: the epoch of the more intensive study of the firmament."

KONSTANTIN E. TSIOLKOVSKY,
Father of Russian Astronautics: 1896

"This nation has tossed its cap over the wall of space, and we have no choice but to follow it."

JOHN F. KENNEDY,
President of United States of America:
November, 1963

Contents

Foreword

The Institute of Professional Engineers New Zealand (IPENZ) Foundation is delighted to have this opportunity to make some remarks in support of this most commendable and authoritative biography of William Pickering by his colleague and friend Douglas Mudgway.

The IPENZ Foundation is a charitable trust formed by the IPENZ in 2002 for the promotion of the engineering profession in New Zealand and to assist in the welfare of its members. William Pickering, one of our most distinguished New Zealanders and an Honorary Fellow of IPENZ, was invited to become the Foundation's inaugural patron. He accepted with alacrity, met with the trustees on his many visits to New Zealand, and maintained a lively interest in the Foundation until his death. He is sadly missed.

In the course of researching the feasibility of sponsoring a biography of William Pickering, the Foundation became aware that preparation of this book was well under way under the auspices of the NASA History Division. We are pleased therefore to be able to perpetuate the memory of William Pickering in New Zealand by our association with this biography, which we see clearly as part of our mission to promote the engineering profession.

William Pickering was a modest man, but his achievements were legion, as the reader will learn from this wonderfully illustrated and very readable biography. He was a spaceflight and rocket engineer and the revered leader of the Jet Propulsion Laboratory at Pasadena in the early heady days of space exploration. The author had the inestimable advantage of knowing Bill and being able to interview him and subsequently, after his death, having full access to his papers.

Douglas Mudgway, also a New Zealander by birth, graduated from the University of New Zealand before moving to the Jet Propulsion Laboratory in 1962 following a 15-year career in Australia in the field of guided missile research.

The IPENZ Foundation is thus doubly proud to be associated with this prestigious NASA publication about a former New Zealander by a former New Zealander. We commend this book to readers in the United States, New Zealand, and around the world who remain in awe of the achievements of the early pioneers of the Space Age. Who can forget the photo (reproduced in the book) of William Pickering, James van Allen, and Wernher von Braun holding aloft the model of Explorer 1 following the successful launch of the first U.S. satellite in 1958?

John Cunningham
Chair
IPENZ Foundation
Wellington, New Zealand

Preface

William Pickering first came to the attention of the world in 1958 when the media triumphantly announced the successful launch of Explorer 1, the American response to the Soviet deployment a few months earlier of the first Earth-orbiting satellite Sputnik. Along with Wernher von Braun and James Van Allen, William Pickering shared the limelight and the accolades. In that instant of time the Space Age was born, and with it the professional reputation of William H. Pickering.

By that time, he had already been the Director of the Jet Propulsion Laboratory for more than three years, and had been associated with the Laboratory for about ten years prior to that time as the head of one of its principal engineering divisions engaged in secret guided missile tests for the U.S. Army.

Shortly after the National Aeronautics and Space Agency (NASA) was established in 1958, Pickering became responsible for carrying out NASA's Ranger program, a bold step to return live, close-up video images of the lunar surface in the last few moments before spacecraft impact. Although the program got off to a discouraging start, Pickering remained confident of ultimate success and, soon enough, the world saw its first close-up pictures of the Moon. These were followed by more sophisticated lunar missions that expanded our knowledge of the Moon and paved the way for the Apollo manned landings on the Moon.

Successful though they were Pickering saw these remarkable achievements as merely the beginning of man's venture into deep space. Under NASA's sponsorship, JPL shifted its focus outward, beyond Earth and the Moon, to the planets, beginning with Venus and Mars. Later, Pickering would push the envelope of JPL's interest even further outward, toward the very edges of the solar system itself, with missions to Jupiter, Saturn, Neptune, and Uranus.

When he retired in 1976, Mariner spacecraft had visited Mercury, Venus, and Mars, and Jupiter had been reconnoitered by each of the two Pioneer spacecraft and two massive Viking spacecraft were in orbit around Mars, each preparing to release a robot Lander to explore the surface of Mars. JPL teams were also preparing to launch two Voyager spacecraft both of which would conduct an amazing 20-year odyssey of all the major planets of the solar system that came to be known as the Grand Tour. This was the legacy that Pickering left for others to build upon, in mankind's relentless pursuit of scientific knowledge and understanding of its place in the "grand scheme of things."

In the years that followed, JPL continued to advance NASA's program of planetary exploration with great success. From time to time Pickering's name appeared in the local newspapers and, those of us at JPL who were interested, learned that he had been invited to Saudi Arabia to setup an institute of technology for the Saudis. A few years later, we heard that he was back in the U.S. and had become involved in development of an alternative fuel for domestic home heating applications. Of the details we knew nothing.

Later, in retirement I found occasion to reflect on what kind of person William Pickering really was and how he had suddenly appeared on the national scene, just when a man like him was needed most. As part of his engineering work force at JPL, I had seen our Director only as a reserved, well-informed man of academic manner whose legendary achievements were a matter of public record, but the persona of this taciturn, tight-lipped man remained hidden from my view.

Early on a brilliant fall afternoon just before Thanksgiving 2002, I called on William Pickering to seek his concurrence and cooperation in writing the story of his professional life. With the passing years, the need to do that had become more imperative and I finally resolved to make the effort. "It might be an interesting idea to kick around," he said in response to my proposal. We chatted back and forth for the rest of the afternoon until it was, obviously, time for me to go. For the next year, in intensive oral interviews, Pickering generously recalled the personal and professional details of his remarkable life, spanning 93 years from childhood in New Zealand to his retirement years in California.

In the following pages I have embedded what he told me in the context of the major events in the American space program in which he played a significant part, significant indeed. In a 1965 article that spoke of Pickering's career at JPL to that time, a leading New York newspaper suggested that his greatest contribution may have been his positive efforts to influence government and public attitudes toward support for the space program, and his determination to rally public confidence in the nation's power to recover from the shock of Soviet dominance in space engendered by the Sputnik affair and subsequent Soviet Moon shots. More than 30 years later, Thomas Everhart, a former president of Caltech, would write, "More than any other individual, Bill Pickering was responsible for America's success in exploring the planets. . . ." These would become his legacies in the American record of space exploration and endeavor.

As in all large enterprises, the top executive gets all the credit despite the obvious fact that the ultimate result is the outcome of the integrated efforts of the thousands of individuals involved. It is also true that the top executive gets all the blame when the outcome turns unfavorable. This was never more true than during Pickering's tenure as Director of JPL. Pickering understood this and thought of himself and JPL, that is, the people of the "Lab" as he called it, as synonymous. Thus, in recalling his story of success and failure, he found it difficult to separate his individual contribution from that of the Laboratory as a whole. The media attention that focused the public spotlight on William Pickering, the individual, tended to overlook the enormous infrastructure that produced the space spectaculars for which he received the credit—or the blame.

Nevertheless, Pickering believed there were two areas for which he was solely responsible. First, he believed that it was his job to create a work environment at the Laboratory that would attract, and retain, the very best engineering and scientific talent to work on its programs. And second, he believed that it was his job to use his public image to foster public support for the U.S. space program and its preeminent position in space exploration. In achieving these ends he engendered strong critics at NASA Headquarters, for his hubris in the former case, and his inordinate expenditure of government time and effort to public speaking and the advancement of professional societies in the latter case. Undeterred by the criticism, Pickering nevertheless forged ahead to realize the ultimate vindication of his responsibilities as he saw them.

After he became Director, William Pickering published little in the way of technical material, preferring rather to make use of his outstanding skills as a public speaker, to present his views and opinions on space and, later, the human condition, to professional and public audiences alike. I have made frequent use of his public speeches to afford a window on his inner thoughts on these topics as they caught his interest over the 20-year period of his involvement with the space program. The archives of both the JPL and Caltech contain much additional material about Pickering and his tenure as Director that remains to be mined by future researchers.

In a life spanning most of the 20th century, William Hayward Pickering rose from the most humble beginnings to achieve worldwide recognition by the highest institutions in the field of science and technology. The institutional story of JPL during Pickering's tenure has been well told elsewhere.[1] This is the personal story of William H. Pickering the man, before, during, and after that climactic period of his life.

Sonoma, California
October 2007

---

[1] Koppes, Clayton R. *JPL and the American Space Program: A History of the Jet Propulsion Laboratory.* New Haven: Yale University Press, 1982.

Acknowledgments

In the task of researching and writing this book, I count myself very fortunate to have enjoyed the personal confidence, support, and encouragement of William H. Pickering for the final three years of his life. After he suddenly passed away in March 2004, I was equally fortunate to have received a continuation of that confidence and support from his daughter, Beth Pickering Mezitt, and his wife, Inez Chapman Pickering. To both of these ladies I am deeply indebted for their gracious help with details of William Pickering's personal life, and for their generous access to the collected personal papers and photographs of William Pickering.

I am also indebted to R. Wayne Mezitt, Trustee for the Pickering Family Trust, for permission to publish the manuscript.

The Archives and Records Section of JPL assisted me with my research in its William H. Pickering Collection and the Millikan Library at Caltech provided me a copy of his thesis and copies of various technical papers dealing with Pickering's early work on cosmic rays at Caltech. The Alexander Turnbull Library in New Zealand generously provided me with background photographs and material related to Wellington in the 1920s to supplement the material on Pickering's years at Wellington College that came from the College Archives Director, Paddianne Neely. The staff at the Marlborough Provincial Museum and Archives in Blenheim, New Zealand, supported this project with background material relating to Havelock for almost 30 years that included young William Pickering's childhood.

A special note of thanks is due to indefatigable Dr. John Campbell of Canterbury University for his concept of the Rutherford-Pickering Memorial in Havelock and his untiring effort to bring the concept to reality. His splendid book on Rutherford provided much insight into life in early Havelock. Also in New Zealand, Alan Hayward and Carol Short, members of the Hayward and Pickering families, respectively, provided recollections and memorabilia for which I am truly grateful.

In reconstructing Pickering's professional career I have drawn upon previously published material from "JPL and the American Space Program," by Clayton R. Koppes; "Exploring the Unknown," by John M. Logsdon; "Beyond the Atmosphere," by Homer Newell; "Rutherford," by John Campbell; "Haywood Heritage," by Stuart Bunn; "Millikan's School," by Judith R. Goodstein; "The Rise of Robert Millikan," by Robert H. Kargon; "The Universal Man," by Michael H. Gorn; and various articles on William H. Pickering from *Time* magazine, *The New York Times*, the *Los Angeles Times*, the *Pasadena Star-News*, Marlborough Express, *The Press Christchurch*, the *Evening Post*, the *Dominion*, and the *Auckland Herald*. I am indebted to the authors and publishers of these various works.

Dr. Steven Dick, NASA Chief Historian, and Erik Conway, JPL Historian, provided insightful comments on my representation of NASA, Caltech, and JPL during Pickering's time for which I am profoundly grateful. Stephen Garber and the staff of the NASA History Division provided invaluable editorial guidance.

In the Communications Support Services Center at NASA Headquarters, Ann-Marie Wildman expertly designed the layout of the book and Steve Bradley adapted the cover art and designed the dust jacket. Stacie Dapoz oversaw the careful copyediting and proofread the layout, David Dixon handled the printing, and Gail Carter-Kane and Cindy Miller supervised the whole production process. These talented professionals gave form and finish to the manuscript, an onerous task that earns my admiration and appreciation..

Finally, I had the unique experience of working as a high-level engineer at JPL during the last 15 years of William Pickering's tenure as Director. His influence on all that happened there in those years percolated down to me, thereby inspiring me and my colleagues alike to do greater things, to reach further, to do better than we or others had done before, to always understand fully what we were doing and why we were doing it, and, above all, to pursue excellence in all that we did. For that experience too, I am grateful.

Chapter 1

The Boy from Havelock

The Monument (2003)

The little town proclaimed its position in the general order of things by two large billboards that were prominently located on the main road at either end of the business area that consisted of several stores, a few cafés, a post office, and two pubs. Against a postcard-like background of green hills, shining water and blue sky, the billboards shouted their greeting: "Welcome to Havelock: Greenshell Mussel Capital of the World." It immediately captured the attention of the tourists and it was good for the town's main business—greenshell mussel farming.

Welcome to Havelock: Greenshell Capital of the World (Photo: S. Mudgway).

However, there was much more to Havelock than these two billboards suggested. In the center of the town, adjacent to the Town Hall stood a tall stone monument known as the "Ronga." Over the years, time and weather had all but obliterated the inscription, and most of the passers-by took little notice of it. The monument commemorated the loss of the schooner "Ronga" in April 1906, in which six local townsfolk lost their lives.[1]

There it stood for almost 100 years as the memories faded and those who remembered passed away. Eventually it became simply an artifact of Havelock, a rather dilapidated symbol of public enthusiasm for a long since forgotten cause.

However, as the new century began, an ambitious initiative began to stir in the town councils of Havelock. It was driven by John Campbell, a professor of physics from the University of Canterbury. In his recent book on the life of Lord Ernest Rutherford,[2] Campbell described Rutherford as ". . . one of the most illustrious scientists the world has ever seen. He is to the atom what Darwin was to evolution, Newton to mechanics, Faraday to electricity, and Einstein to relativity."[3] Although Rutherford was born in nearby Nelson, he received his primary education at the tiny country school of Havelock.

In John Campbell's mind, the New Zealand public paid insufficient homage to its world-famous son, and he was determined to do something about it. For a decade, he worked to promote public appreciation of Rutherford by giving talks and lectures in schools and conferences, promoting exhibitions and displays in public places, and distributing information to schools. By 1991, he had been instrumental in implementing the "Rutherford Birthplace Project" a memorial plaza in the adjacent city of Nelson, where Rutherford went to secondary school, or "college," as it is called in New Zealand.

However, the splendid Rutherford Plaza, being in Nelson, left the town of Havelock without any formal recognition of its association with the great man. Campbell could not resist a challenge to do something about that, too.

At that point, Campbell recalled meeting another famous New Zealand scientist during his visit to the University at Christchurch in the early 1980s. In the course of their conversations, Campbell learned that, by an extraordinary coincidence, this famous scientist had also lived in Havelock and been educated in the same little primary school at Havelock, just a few years after Rutherford. Like Rutherford before him, he had to move elsewhere to complete "college" and university education before going on to attain world fame in another country. His name was William Pickering and he went on to become the pioneer of America's space exploration program.

This was the compelling reason Campbell needed. He would give Havelock a memorial to not one famous son, but two: Earnest Rutherford and William Pickering—both giants in stature on the world scene of science and technology. Campbell persuaded local government officials that a memorial plaza to recognize Havelock's association with these two famous scientists would enhance the status of the township. It would not degrade its natural beauty and, he said, "It would probably increase Havelock's tourist trade." The "Ronga" monument would be a perfect site for the new memorial. With this agreement in hand, Campbell initiated the project.

On 15 March 2003, in the presence of representatives of the principal patrons, civic officials, dignitaries from the local Maori tribe, Ngati Kuia,[4] an assembly of school children, some 200 guests and many curious tourists, William Pickering, the surviving honoree, unveiled the Rutherford-Pickering Memorial. It stood adjacent to the "Ronga" monument, near the town hall on the main road.

For New Zealanders, their connection with an international figure of the stature of William Pickering was cause for considerable national pride. His achievements were legendary. In 1958, he had led America's successful challenge to the Soviets' bid for technological supremacy.[5] His image had appeared on the cover of *Time* magazine on two occasions, first in 1962 for the world's first robotic spacecraft visit to Venus, and again in 1964 for an encore expedition to Mars. These missions were not of course, individual ventures, they were part of the United States' space program, which was managed and executed by the National Aeronautics and Space Administration (NASA). They were led, however, by individuals that possessed the unique technological experience and who had innovative minds, driving motivation, and confidence to go where none had gone before, and whose reputation and personality

The Rutherford-Pickering Memorial Plaza, Havelock, March, 2003. The "Ronga" monument stands to the center rear of the photo. The historic town hall is to the left-Mussel Boys Café is further down the street to the right (Photo: D. Mudgway).

inspired others to follow. Those were necessary, but not sufficient conditions for success. Being in the "right place at the right time" was the culminating and sufficient condition. Such a man was William Hayward Pickering, a quiet boy from Havelock who rose to become a distinguished public figure, showered with international honors for his contributions to the advancement of human knowledge in the esoteric world of scientific space exploration. Here was where it all began.

Havelock (1913)

William Pickering (senior) came from a sturdy line of English immigrants who had settled originally in Australia in the mid-1800s before immigrating to New Zealand in 1860. He established a coaching business in 1879 and for many years thereafter, William Pickering's coach was a familiar sight and service to travelers on the roads between Havelock and Blenheim.

In 1880 at age 34, William Pickering, Coach Proprietor, by then a successful and well-liked citizen of Havelock, met and married the beautiful Miss Kate Douslin; she was the daughter of William Douslin, a prominent member of the Blenheim Borough Council.

Mr. W. Pickering's Coach: First to make the journey from Blenheim to Nelson in 1885 (Photo: Carol Short).

As the years passed, William and Kate were prosperous. They would never become wealthy, but they were solid pioneering citizens struggling to make a life under the harsh conditions common in the remote areas of early New Zealand. Although both William and Kate were folk of little education, they appreciated the value of learning. All six children were enrolled in succession at the Havelock Primary School. At any one time in the mid-1880s, there was a Pickering child in every class from the Lower School to the Upper School. "Pickering" became a most common name in and around the Havelock area. Education was encouraged in the Pickering household and all of the children did well at their schoolwork, but their first-born, a son named Albert, did particularly well.

By the time Albert Pickering reached the age of six and was ready to start school, an older boy named Earnest Rutherford had already been at the upper classes of the school for two years and was about to enter the examinations required to continue his education at one of several prestigious secondary schools or "colleges."

Albert Pickering followed a similar path through the higher educational system of the times, eventually earning a diploma in Pharmacology. He embarked on a professional career as a qualified pharmacist, and married Elizabeth Ann Hayward, an attractive businesswoman from the Dunedin area of New Zealand in 1908.[6]

Shortly thereafter, the newlywed couple moved to the capital city of Wellington, where Albert "Bert" took up his business profession in the city, while Elizabeth Ann "Bess" set up a home and prepared for the birth of their first child. William Hayward Pickering arrived on Christmas Eve 1910, and was regarded by both adoring parents as the ultimate Christmas present. Their joy and happiness was shared by both sets of grandparents and their large families of brothers and sisters in Havelock and Dunedin.

A second child, a son they named Balfour, was born in 1913 and their family seemed complete. By then, Bert had entered service with the New Zealand government as a pharmacist in the Office of Public Health. Their prospects were bright and future seemed secure. Then tragedy struck.

Shortly after the birth of Balfour, Bess contracted peritonitis, and died in June 1915 after a protracted period of severe illness. She was 38 years old. For a while, Bert struggled to hold the home together while he tried to reshape his life after the loss of Bess. That, however, was not to be.

A few months later, baby Balfour contracted diphtheria and despite Bert's professional expertise, the drugs available then could not save the child's life.

To exacerbate the problems facing Bert Pickering, the Office of Public Health assigned Bert to an overseas position as Public Health Officer at Apia, Western Samoa, an area controlled by Germany prior to World War I; now occupied by New Zealand troops. Realizing that it was impossible to take young Will with him under the circumstances, Bert made an agonizing decision to part with the sole remaining member of his family. He transferred the care of young Will to his grandparents in Havelock, then moved from Wellington to take up the government position in Samoa.

The home of William and Kate Pickering, humble though it was, provided a loving and accepting environment for a young child and, after a short period of anguish at the separation from his father, Will settled down and soon adapted to the new world that surrounded him.

From time to time, Will's father returned to New Zealand on furlough from Samoa. On one such occasion in December 1916, Bert brought presents for his son's 6th birthday and enjoyed a Havelock Christmas with his family. The new year came and passed rather somberly as New Zealand marked the third year of its engagement in World War I while Bert stayed on in Havelock partly to be with his son, but more particularly to start Will at school. It seemed important to Bert that he should see Will off to this important new phase of his young life.

In the New Zealand education system of the time, children began their primary school education at age 6 and remained in that system until age 12 or 13, after which they could, if they qualified by examination, continue with their education at the secondary school level. The Havelock Primary School was part of this system.[7]

Will Pickering soon made an impression at Havelock Primary School. Well-behaved, quick to learn, interested in everything, and equipped with a naturally retentive memory, Will quickly grasped the basics of reading, writing, and arithmetic. Always a great explainer, Will soon began to recreate school at home, where he would play the role of teacher while his amused grandparents played his classmates.

Two years after he began, Will Pickering passed out of the Lower School into the tutelage of a Mr. Barraclough, in the Upper School. At age eight, he regarded himself as a big boy, and with his customary quiet confidence began the new school year of 1919 at the class level of Standard 2. It was not long before Mr. Barraclough too, began to notice the presence of an outstanding pupil among his new class, and began to take a particular interest in his progress. Will became known as the "smart" boy of the school.

Year after year, Will moved steadily through the ever-advancing class levels, demonstrating superior skills in all subjects but excelling particularly in science and arithmetic. Will's scholastic ability was such that he was able to combine two of the standard levels into a single year, and to expand the breadth of his studies to include algebra and Latin in addition to the regular curriculum of English, composition, history, geography, and science. He excelled at all of them.

Although Bert seldom saw his rapidly growing son, he was able visit Havelock on several occasions when on leave from his government post in Samoa. He would have been astonished each time to see the increase in Will's stature, self-confidence, and general knowledge. They went on outings together, and on more than one occasion visited the Hayward family in Dunedin.

As Will's reading and writing skills rapidly matured, Bert was able to keep in touch by exchanging letters with Will, and to reassert his presence in the child's life. Will's letters described the more significant things that were happening in his young life.[8]

From time to time during the school holidays, his mother's family, the Haywards, invited Will to stay at their large family home in Dunedin. Those times were filled with great fun and affection, and Will fit into the large family environment. The younger son, Jock, was about his own age and the two boys enjoyed each other's company, and eventually became life-long friends.

At that time, Havelock, like most other small towns in New Zealand, did not have electricity. However, a simple electric generator driven by water from a small dam nearby provided current for about two hours each evening for those few citizens affluent enough to have installed electric light. This literally "sparked" the interest of young Will in a way that would become evident in the years ahead.

Will Pickering turned 11 years old in December 1921, and looked forward to entering the final class, Standard 6, at Havelock when school resumed in February 1922 after the long Christmas vacation. His confidence was made evident by his class marks; he was an outstanding pupil and had always been at the top of his classes.

In the New Zealand educational system at that time, a Certificate of Proficiency was required both as evidence of a completed primary education,

Albert William Pickering "Bert" with son, William Hayward Pickering "Will," circa 1919 (Photo: NASA/JPL-Caltech Archives P2395B).

Will Pickering strikes a wild game hunter's pose at the Hayward's holiday cottage near Dunedin, circa 1921 (Photo: NASA/JPL-Caltech Archives P2396A).

and as a necessary qualification for continuation to a secondary education in either a high school or a college. It was an essential key to the future education of all children throughout the country. The dreaded proficiency examination was administered at the end of the year in which the pupil reached the Standard 6 level, and carried with it a certain aura of apprehension among children and parents alike. The Pickering family, however, had no concerns about Will's ability to pass the proficiency exam, and never doubted that he would go on to college and do well. They simply assumed that he would succeed. And of course he did, with top grades.

When the proficiency test results came out early in the new year, Will's name topped the list of those who passed. He received a Certificate of Proficiency to mark his success, and a short time later, his father arrived back from Samoa to take him to college[9] in Wellington.

Will Pickering would not have known it at the time, but he was about to step into an environment that was the very antithesis of the gentle, loving, and caring atmosphere that had surrounded him in Havelock, but one which would lead him in due course to his destiny on the other side of the world. Will Pickering was, as they would say in New Zealand, "going to college."

Wellington (1923)

In February 1923, when Bert Pickering enrolled his 12-year-old son at prestigious Wellington College, the college had stood on its present site for almost 50 years. For a new country like New Zealand, the college had already established a long and distinguished history.[10]

The school was steeped in tradition, and prowess in the fields of art, scholarship, government service, military service, and sport: cricket and rugby were most highly regarded and memorialized in the school's list of honorees and honors boards. Those who succeeded in these areas were held up as heroes, and lauded as examples of selfless dedication and loyalty to school, country, and Empire. In short, the school exemplified the spirit of the time. It was a strong reflection of New Zealand's British origin and its close ties to England and the British Crown. Discipline was strict, corporal punishment was accepted, military training was part of the regular curriculum, and scholastic standards—measured by frequent, unforgiving testing—were high.

Wellington College adhered strictly to the common English public school practice of a uniform dress code. All pupils, regardless of age, were required to wear the school uniform at all times during school hours and during travel to and from school. There were penalties for infringement of the dress code rules. The official uniform, which was quite an expensive item for those times comprised short, gray, knee-length pants, a gray tailored jacket, gray woolen shirt and tie, knee-length pull-up black socks with two gold bands—which were the school colors—around the tops, black dress shoes, and black peaked cap with radial gold stripes. At the front, embroidered in gold thread, the cap bore a medallion proclaiming the school icon, an oil-lamp surmounted by a banner bearing the school motto in Latin, *Lumen Accipe Et Imperti*: "Receive the Light and Pass it On."

Pupils of Wellington College, circa 1923. In front row are Lord Jellicoe (center) and Headmaster J. P. Firth (left). The school buildings are in the background (Photo: Alexander Turnbull Library, Wellington, New Zealand, Image No. 005413).

Also in keeping with English public school practice, the school year was divided into three terms, with a long break of six weeks following the third term that corresponded to summer; Christmas and New Year in the Southern Hemisphere. The first term began in February and the third term ended in December.

When the first term of the college school year began in late summer 1923, the school roll numbered about 800, of which about 60 were boarders. Will Pickering was among the annual intake of new boys, and he was a boarder, enrolled on the "general" side. Henceforth he was known formally as Pickering, W. H., or just plain "Pickering" among his masters and peers. To the other boarders he was "the boy from Havelock."

Although Pickering's father remained in Samoa the entire time he was at college, his son seemed to be quite happy with the arrangement, and satisfied to divide his vacation time between Havelock, Dunedin, and his Uncle Horace Douslin, at Rotorua.[11] As "Willie," he was an excellent correspondent, and wrote regularly to both his father and his grandmother describing the events of his new life at college.[12]

For a boy who had just turned 12 years old, Pickering was rather tall, of slight build, and was healthy and strong. He was shy, with fair hair, and a quiet and thoughtful disposition. Although he was friendly enough to those with whom he came in contact, he was not particularly gregarious, and seems to have made few close friendships during his college years. But he was adaptable, self-confident and smart, as events soon showed.

Pickering quickly adapted to the school and boarding house routines without suffering psychological stress due to separation from his Havelock home and family. Undoubtedly, his innate academic ability eased the burden of a full load of class work: English, French, Latin, mathematics, science, history, and geography. His regular exam results were consistently at the top of his class and in the masters' "common room," the "boy from Havelock" had begun to attract attention as a quiet achiever who would ultimately[13] do well.

In addition to keeping up with his class work, Pickering involved himself in several non-academic activities in that first year.

He had learned how to shoot a .22 gauge rifle during his past holidays with the Haywards in Dunedin and brief as it was, this prior experience quickly led him to a place on the college shooting team. He joined the debating society and wrote: "Last Thursday we had a debate on the subject of coeducation in secondary schools. It was defeated"; played football: "I have been playing football this term, also taking dancing [lessons]"; and achieved some success in athletics: "I won by about a yard or so in 5 minutes, 22 seconds."[14]

However, his interest in another activity, the radio club, was to have far-reaching significance for him in ways that he could not possibly have imagined. The first glimmer of interest appeared in his mid-April letter to his father, "On the 1st, there was a meeting of the radio club. The class instruction on the theory of wireless commences next Thursday."[15]

In 1923 when Pickering started college, the "wireless" was just beginning to make its public appearance in New Zealand as a new entertainment medium. Already, amateur radio enthusiasts had established an active, worldwide organization and, by 1923, there were a substantial number of amateur radio stations operated by "hams" scattered throughout New Zealand. They exchanged messages between other amateurs in Australia, the U.S., and other countries around the world, as their limited equipment would permit.

Wellington College was the first public school in New Zealand to have its own amateur radio station and it was the radio club that initially attracted Pickering's interest. In Pickering's first year, the club had built wireless receiving equipment which was used to detect and log amateur broadcasts from all over New Zealand and, occasionally, from Australia and the U.S. depending on conditions. He asked his father to buy him a wireless valve (radio tube)—a very exotic device at that time—to build an improved wireless receiver, "I think the best kind of valve will be a V201A. I have a single valve set working here and I need a transformer and another valve to make a two-valve set out of it. This should be able to get America and perhaps, England." However, the club did not have transmitting equipment, nor did it have a license to operate a transmitter. That would come later.

By the time the 1924 school year began, Pickering, W. H. was, in many respects, a different boy. Now aged 13, he had grown a full 3 inches and had gained 14 pounds in weight. No longer a "new boy," he had moved into the A class of the fourth-form level. Quite secure in the college environment, he excelled at his class work and again engaged in many of the extra-curricular activities that were available to him.

The school curriculum included compulsory military training, and on Friday afternoons, the entire school turned out for a formal parade followed by officers' inspection, marching, and rifle drill.

He signed up for boxing classes, "My boxing has been alright. I have not been hurt much, yet."[17] However, it was the activities of the radio club that continued to attract his greatest attention. The year 1924 had been an important one for amateur radio at Wellington College. The club already had a wireless receiver, but that year the club members constructed a wireless transmitter, and several members acquired the necessary license to operate it on the international amateur radio bands. Pickering was caught up in the excitement of building the transmitter, putting radio Z-2BL on the air and using it effectively to contact other amateur stations throughout New Zealand and Australia. This early practical experience with radio propagation and electrical circuits was to have a seminal influence on the course of his career in later life.

His father's ideas for his son's future career reflected these emerging interests. William wrote his grandmother: "Yesterday I had a letter from Dad. He says that he thinks that a position in the Pacific Cable Company would be a good one for me. I think so too."[18]

Cadet Pickering W.H., Wellington College Cadet Corps: 1924 (Photo: NASA/JPL-Caltech Archives Laboratory 75-4).

Another school organization that caught Pickering's attention—the Natural Science Society—was run by a member of the school staff, Mr. Stevens, who gave regular talks on topics that fell marginally within that category. Topics such as "optical illusions," "fishing," "chemical magic," "electro motors," and "the lighter side of physics" were typical. However, the *Wellingtonian*, the college yearbook,[19] reported that the best lecture for that year was the one on "explosives." It was a standing-room only audience and later many of the pupils "indulged their love of noise with a good deal more enthusiasm than discretion." Pickering thought so too, "I went to a lecture on explosives . . . that went off on their own. A good mixture was made of sulphuric [*sic*] acid, nitric acid, and turpentine. On Saturday I bought the things but they would not work. I found out what was the matter and I will fix it and try again on Saturday" he told his father.[20] Young Pickering soon learned how to make percussion detonators that could be exploded by stamping on them, or striking them with a hammer much to the delight of his young friends.

In 1925, Pickering advanced to the fifth form in a special class for gifted pupils. Chemistry, mathematics, English, and Latin were the principal subjects, taught at an advanced level in order to prepare the boys for two major scholastic challenges, the Matriculation, and the Senior National Scholarship. "At the end of the year, I am entering for the Senior National Scholarship. This gives me £45 a year for two years. At present I am getting £40 a year for three years" he reported to his father.[21] Both were key requirements for their academic future, and both were perceived by the school as a measure of its academic standing in New Zealand's educational system.

When the radio club was reorganized in 1925, Pickering was elected to the committee and began to influence the club's activities. He obtained his Amateur Radio Operator certificate, and set about improving the transmitter, receiver, and antenna. The station immediately began making contact with stations in the U.S. and an increasing number of Australian stations.[22]

Pickering turned 15 in December of 1925 and spent the 6-week summer holidays with his Uncle Horace Douslin[23] on the dairy farm near Lake Rotorua in the center of the North Island. Horace Douslin was the brother of Kate Pickering, and was technically Bert's uncle. However, within the family, Will Pickering always knew him as Uncle Horace. Now retired from a professional career in Rhodesia, Horace was unmarried and comfortably well off, and had purchased the undeveloped property at Rotorua a few years earlier. Through occasional meetings in Havelock, he had formed quite an attachment to this bright Pickering lad. For his part, William Pickering was a welcome and helpful addition to the daily routine of farming life in rural New Zealand.

The Matriculation Certificate from the previous year had marked the end of Pickering's general secondary education. Now in the lower sixth form, his academic courses, English, mathematics, and science, began to prepare him for continuing on to a tertiary education at the university.

His remarkable aptitude for science and mathematics had already brought him to the attention of a senior member of the school staff by the name of A. C. Gifford. During his long association with the school, Gifford had established a small astronomical observatory equipped with a fine, 5-inch, refractor telescope on the school grounds. Being a first-class graduate from the University of Cambridge, England, and a Fellow of the Royal Astronomical Society of England, Gifford was well qualified in the science of astronomy. With his gentle disposition and fatherly manner, he attracted boys like Pickering and encouraged them to learn how to use the telescope for making astronomical observations. Gifford's instrument gave Pickering his first views of the Moon, Venus, Mars, and Jupiter. He was spellbound at what he saw. The Gifford Observatory became an additional source of absorbing interest for him in those years at college, and established within him an abiding interest to learn more about our solar system.

Pickering began his final year at Wellington College in February 1927. At 16 years of age, he had acquired an enviable school record. Supported by a network of loving people who cared greatly about him, Pickering thrived on the challenges that school life brought to him. Although the complexity of his schoolwork now approached first year university level, he appeared able to cope with the extra demands it made upon him without undue strain, and the final year passed quickly.

Under Pickering's leadership, the radio club converted a 3-valve Browning-Drake receiver to a 5-valve set, and dazzled the school assembly with a high tech demonstration of "radio reception on a loudspeaker using an indoor aerial," a major technological advance for the time.

He sustained his deep interest in the Observatory and, with Gifford's encouragement, used it to broaden his knowledge and become familiar with the motions of the Moon and planets, along with their surface features and general characteristics. But he was not satisfied. A fine instrument though it was, he could never see enough through the Gifford telescope, and always wished he could see more. If only he could get closer—but that was just wishful thinking, or so he thought.

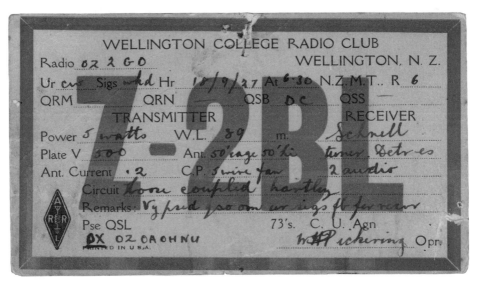

QSL Card (Acknowledgement of radio contact) dated 18 September 1927 from Z-2BL from Wellington College radio club signed by W. H. Pickering, Operator (Photo: Wellington College Archives).

As a member of the Upper Sixth Form, the top class in the school, and an outstanding pupil in addition, Pickering was then regarded as person of distinction among the general student body.

In June of his final year at college, Pickering received a letter from Horace Douslin that was to set his course for the future. Pickering passed it on to his father: "Enclosed, is a letter from Uncle Horace and one from Grandma. You see what he is offering me. So far as I can see, it ought to be a very good thing. . . . Perhaps it would be better to get [my] B.E. here, and then go on [to America]," he wrote. He sought some advice at the college: "The other day I saw Mr. L. He seems all in favor of my going to America, but says we can't do much until we find out which university it is. He says there is a good one in Pasadena about fifteen miles from Glendale."[24]

Soon enough they would come to find out that Horace Douslin had been speaking of the California Institute of Technology.

Undistracted by the uncertainty associated with a future university career in America, William Pickering pressed on with his school activities for the final year in Wellington. However, the American idea eventually fell through and, he felt that his immediate goal should be to complete an engineering degree at Canterbury College in Christchurch, and so to that end he set his sights.

By the end of the year, he had added a University National Scholarship to his personal list of secondary school achievements and in late December at the age of 17, left Firth House for the last time and returned to Havelock for a short summer holiday.

Wellington College Prefect, 1927
(Photo: Wellington College Archives).

The little town was different now; electricity had arrived and most of the houses, including the Pickering's, now had electric light whenever they needed it. Grandmother Kate never ceased to marvel at the wonder of her electric light switch. The single main street was lighted and the road through the town had been paved. Automobiles had arrived in Havelock and, although a few horse-drawn vehicles were still a common sight in the town, the motorcar was rapidly replacing them. Petrol arrived in large cans, two to a box. There was a regular moving picture show every Saturday night in the town hall. The "talkies" had not yet come to Havelock, but they never missed what they never had, and all were content.

It was, in a sense, the end of his age of innocence and the end of his boyhood. He had stepped out alone, looked at what the world outside Havelock had to offer, and decided for himself what he wanted to do next. For him, the learning process had only just begun, and university—the road to higher learning and discovery—beckoned him.

Later that summer he moved to the large, South Island city of Christchurch, with the intention of enrolling at Canterbury University College, the engineering school of the University of New Zealand.

Christchurch (1927)

The city of Christchurch took its name from Christ Church College, Oxford, England, some of whose members, led by the Archbishop of Canterbury, formed the Canterbury Association in 1849. They intended to establish a middle-class Anglican community in New Zealand where the moral values of Victorian

Endnotes

1 Marlborough Provincial Museum and Archives, Brayshaw Museum Park, Blenheim, New Zealand, March 2003.

2 Campbell, John. *Rutherford: Scientist Supreme* (Christchurch, New Zealand: AAS Publication, 1999).

3 Ibid.

4 Official party from the neighboring Maori organizations.

5 Sputnik, the world's first Earth–orbiting satellite was launched by the Soviets in October 1957.

6 Bunn, Stuart. *Hayward Heritage: The Story of a Pioneer Family* (Mosgiel, New Zealand: Stuart Bunn, 1996).

7 Congdon, Eldred. *A Century of Education in Havelock* (Blenheim, New Zealand: Marlborough Historical Society, 1961).

8 Pickering, W. H. Letter to Mr. I. J. Horton, "Havelock School Centennial Celebration, 25 August 1961." Marlborough Historical Society, March 2003.

9 In general educational terms, "college" was equivalent to "high school" in the United States. It is to be distinguished from U.S. common usage of "college," meaning University.

10 Beasley, A. W. *The Light Accepted: 125 Years of Wellington College* (New Zealand: Wellington College, 1992).

11 Mezitt, Beth Pickering. Private correspondence with the author, May 2003.

12 Most of the letters he wrote during his college years have been preserved, and are referenced in what follows under "Letters" in the bibliography.

13 Pickering , W. H. Collected Letters of William Pickering to his Father: 1923—1927. Beth Pickering Mezitt in correspondence with the author, 2004.

14 Ibid.

15 Ibid.

16 Ibid.

17 Ibid.

18 Ibid.

19 The *Wellingtonian* was the college yearbook.

20 Pickering, W. H. Collected Letters of William Pickering to his Father: 1923—1927. Beth Pickering Mezitt in correspondence with the author, 2004.

21 One New Zealand pound (£) was equivalent to about five U.S. dollars ($) at the time; also see Letters, 2004, in the Bibliography for more information.

22 Wellington College Archives. The *Wellingtonian*. Wellington, New Zealand: Vol. 34, Wellington, December 1925.

23 Horace Douslin was the brother of Kate, and technically Bert's uncle. Within the family, Will Pickering always knew him as Uncle Horace.

24 Pickering, W. H. Collected Letters of William Pickering to his Father: 1923—1927. Beth Pickering Mezitt in correspondence with the author, 2004.

25 The first of the colleges was Victoria College in Wellington.

26 Campbell, John. *Rutherford: Scientist Supreme* (Christchurch, New Zealand: American Astronautical Society Publication, 1999).

27 University of Canterbury Archives, 2003.

Chapter 2

The Cosmic Ray Researcher

Pasadena (1929)

Eighteen days after leaving Wellington, the *Makura* arrived off the coast of California, paused briefly to pick up a pilot, and then slipped quietly through the unspanned towering Marin Headlands of the Golden Gate to a dock near the foot of Market Street, San Francisco. It was early March 1929.

The Douslins had arranged for a car and driver to take them south to Los Angeles, a leisurely drive of some 400 miles that they thought would provide an interesting and instructive experience for their young guest from New Zealand. A week later, travel-weary and quite sufficiently impressed with the sweeping grandeur of the California landscape, the party arrived in Glendale, an outer suburb of Los Angeles.[1]

The Douslins had arranged accommodation for William in an apartment house they owned near Beverley Boulevard, a middle-class locality to the west of downtown Los Angeles. He would have easy access to the elaborate network of electric rail transportation that serviced the Los Angeles area at the time, and he could live there until he found more convenient accommodations closer to the California Institute of Technology that occupied a large campus on the outskirts of Pasadena, about an hour's tram ride away.

From its humble beginnings as Throop Polytechnic Institute in the early part of the century, Caltech matured steadily under the inspired leadership of Hale, Noyes, and Millikan to reach maturity in the 1930s. Speaking of Caltech's status in the world of science, historian Judith Goodstein wrote, "Albert Einstein's visits to the Campus in 1931, 1932, and 1933 capped Millikan's campaign to make Caltech one of the physics capitals of the world."[2] However, dominant though it was, physics was not the only world-class research in progress at Caltech in the 1930s.

In 1932, Charles Lauritson's pioneering work with particle accelerators marked the beginning of nuclear physics at Caltech. Linus Pauling was engaged in studies of the chemical bonds of molecules, a field that would earn him a Nobel Prize for chemistry in 1954. Charles Richter was investigating earthquake phenomena from which the universal Richter Scale for defining the magnitude of earthquakes would emerge. Under the brilliant direction of Theodore von Kármán, Caltech's Guggenheim Aeronautical Laboratory was using its wind tunnels to assist the aircraft industry with the development of improved, safer commercial aircraft such as the Douglas DC-1, DC-2, and the famous DC-3. Working with the Mt. Wilson telescope a few years earlier, Caltech astronomer Edwin Hubble had discovered the red shift spectra of galaxies and linked his observations to the concept of an expanding universe. In 1936, to great public acclaim, the huge Pyrex glass disk that would eventually become the reflector for the 200-inch Hale telescope on Mt. Palomar arrived at Caltech to begin the long, tedious, and exacting task of shaping and polishing to its final exquisite shape. At Mt. Palomar, the giant framework that would house the reflector and become "the perfect machine," was being assembled to meet the specifications of Caltech engineers and astronomers.[3] And then, presiding over it all, there was Robert Millikan.

Such was the academic environment in which William Pickering chose to shape his future at Caltech in the decade of the 1930s. There were few places on Earth where such an array of scientific talent was focused in one place at one time. From the 1920s at Wellington College to the 1930s at Caltech was indeed a giant leap that, for any lesser man than William Pickering, would have posed a daunting challenge.

The school year had already begun when William arrived in Pasadena, and he was well aware that his chances of acceptance for the current year would be further diminished by any delay in submitting his application for enrollment. Nevertheless, he was determined to deal with the problem immediately. A short time later, he presented himself and his school reports from New Zealand to the student admissions office at Caltech to apply for enrollment in the electrical engineering course for the graduating class of 1932.

At that time, the Caltech school year was based on a three-term system, beginning in September and ending in June. Each term was of three months duration, and students were required to take examinations at the end of each term as a prerequisite to continuing on to the following terms. By the end of March 1929, when Pickering made his application for enrollment, first year engineering students at Caltech had already completed their second term exams for the school year that began in September 1928, and were about to commence their third and final term.

Somewhat nonplussed by the unexpected arrival of a young student from the antipodes, but nevertheless impressed with his school credentials, the Caltech authorities condescendingly offered to allow him to take the second term exams, the outcome of which would determine their response to his request for admission. If they were somewhat surprised at his ready acceptance of their offer, then they were completely astonished at the high quality of his examination returns and they readily approved his admission to the engineering class of 1932. In April 1929, William H. Pickering became an official member of the freshman engineering class of 1932 at California Institute of Technology.

The third and final term classes began almost immediately and left William no time to seek more convenient accommodations nearer Caltech. Nevertheless, he quickly settled into a routine and, despite his problems with travel and time-consuming tasks of providing for himself, he was able to complete the school year with very satisfactory grades.

By then, he had found new accommodations in Pasadena, closer to Caltech, and became friendly with his housemate and fellow student, Gordon Bowler. They often went on outings together and shared a mutual interest in outdoor activities.

While the course work in the freshman year had been general in nature and common to all engineering students, the course work in the sophomore year became more oriented toward the electrical engineering curriculum. None of it, however, presented William with any problems. Rather, he found the course work an interesting challenge and the environment in which he now found himself not unlike what he had adapted to at Wellington College. In fact, he did so well in his term exams that toward the end of the year he attracted the attention of Caltech president, Robert A. Millikan. After a short but no doubt intensive personal lecture in the great man's study, he persuaded the young student to change his ultimate goal from electrical engineering to physics. Although Millikan's motives for his interest in Pickering were not obvious at the time and, in retrospect, may well have been self-serving, the consequence of Pickering's decision to follow his mentor's advice was to have a profound effect on the course of his subsequent career.

Pickering's sophomore year had ended on a high note. After his talk with Millikan and his decision to change his major from electrical engineering to physics he must have felt very satisfied with his new life in America, and looked forward to a relaxing summer once again.

In December 1930, William turned 20 years of age. No longer a newcomer, he was well established at Caltech and had begun to cultivate a circle of friends who, like himself, enjoyed hiking the nearby San Gabriel Mountains. He particularly enjoyed the company of Gordon Bowler on these excursions, particularly when Gordon brought his sister Muriel along on one or two occasions. He had already achieved a solid academic record and was embarking on a course that would lead him to a bachelor of science (B.S.) in physics the following year.

That year he won the Caltech Junior Travel Prize. The prize was awarded every year and provided money ($900 per person) for two juniors to spend six months traveling in Europe. The winners for 1932 were William Pickering and Charles Jones.

Before departing, Pickering and Jones were joined by two other students who traveled at their own expense. Being already somewhat of a worldly traveler, Pickering naturally assumed the role of leader. Under his persuasive direction, the group "circumnavigated" Europe. "It was," said Pickering many years later, "really a boondoggle for a bunch of college kids."

Late in 1931, the Caltech administration completed construction of several on-campus residential units, or houses, for its undergraduate students. It was believed that the social interaction resulting from this form of accommodation would enhance the students' university experience, and give them a more "rounded" education to take in to the world beyond academia.[4] Pickering took advantage of this opportunity, more for convenience than for rounding out his education, and moved into the "Dabney" residence on campus along with his friend Gordon Bowler—just in time for the start of his final undergraduate school year.[5] The course work, physics, mathematics, or chemistry, continued to absorb his interest and he excelled in the grade ratings. He was elected senior class president and began to take a serious interest in Muriel, Gordon Bowler's attractive sister. Muriel had become a professional librarian and worked at the Echo Park Branch of the Los Angeles Central Library. Muriel was fascinated by this quiet, courteous, and very smart young Caltech student that spoke with a most unusual accent.

At the end of the school year, Pickering took the final exams, passed with honors, and graduated with a B.S. in physics (1932). The Douslins came to the graduation ceremony to wish him well and so did Muriel. He wrote his father with the good news and mentioned that he had met a fine young woman in whom he had found a great deal of common interest and whose company he enjoyed very much. The summer that followed afforded him the opportunity to see more of Muriel. Together they made hiking trips in the San Gabriel Mountains, spent time at the beach, made day trips to Catalina Island. Their relationship blossomed while they enjoyed the theater, movies, and social outings.

Soon enough the summer passed and in September 1932 Pickering was hard at work again on the first-term course for his master's degree. He had applied for and won a graduate fellowship that, together with some financial help from Horace Douslin, enabled him to get by with his living expenses. During the short Christmas/New Year break at the end of the year, he marked his 22nd birthday. Six days later on 30 December 1932, in a small private chapel at Forest Lawn, Glendale, California, he married Muriel Bowler. After a short honeymoon, the couple moved into a small, but comfortable, apartment on the edge of the lake in the Echo Park district of Los Angeles, close to Muriel's library. 1932 was a busy year for William Pickering.

The following year passed rapidly as William learned to make the personal adjustments that accompany the transition from single to married life. To all appearances he was able to deal with those personal affairs and maintain a high level of achievement with his course work for at the end of the school year he once again graduated with honors and was awarded the degree of master of science in physics (1933). As soon as the graduation ceremony was over, he and Muriel headed for a hiking trip in the high Sierras. Refreshed in body, mind, and spirit by the pristine surroundings and the crystal-clear air of the high country, Pickering began to consider his options for post-graduate work for his doctoral thesis.

Largely as a result of Millikan's advice two years earlier, he had switched his major from electrical engineering to physics. Now, in need of a topic for his doctoral thesis, he turned to Millikan for advice. To Millikan the answer was clearly obvious—it was "cosmic rays."

Now in his sixth year at Caltech, Pickering did what most graduate students did in those days, he sought a job to help pay his expenses and began his first research program. He had secured a Coffin fellowship that provided some financial support, and he supplemented this with a part-time teaching appointment for the undergraduate classes.

It would, he expected, take about three years to complete the research and write his thesis. Since Millikan had taken an early interest in his academic career, Pickering elected to join his cosmic-ray group. It was a measure of his potential that the world-famous scientist accepted Pickering into his circle of elite scientists. He would be one of three researchers. Carl Anderson, later to become a Nobel Prize winner, ran the cloud-chamber experiments; Vic Neher ran the electroscope experiments which Millikan himself had initiated in previous years; and Pickering would employ the recently developed Geiger tubes for his measurements. The young man from New Zealand was in heady company.

Muriel Bowler and William Pickering Pasadena, 1933 (Photo: Courtesy of Pickering Family Trust).

Cosmic Ray Researcher

In an address to the National Academy of Sciences in November 1925, Robert Millikan reported on recent studies at the California Institute of Technology of what he chose to call "cosmic rays."[6] The descriptive name he bestowed upon this high-energy radiation derived from his belief that they were of cosmic origin, that they rained upon Earth from sources within the Milky Way or beyond in the unfathomable reaches of the cosmos. He reasoned that the rays were produced as a result of some kind of nuclear transformation whose energy was much greater than anything hitherto observed in radioactive processes on Earth, and that the cosmic rays were indicators of these changes. Perhaps, he surmised, these processes resulted from the conversion of hydrogen nuclei to helium atoms, or the transformation of some other atomic nucleus from one form to another. In his excitement he envisioned cosmic rays as the "birth cries of infant atoms" born out of a process of fusion or electron capture. On a later occasion he referred to them as the "music of the spheres."

In the early 1930s, Caltech, Millikan, and cosmic rays were inseparably associated with the origins of the universe in the public view of American science.

However, by the mid 1930s, Millikan's original and much publicized "birth cries of the elements" hypothesis regarding the origin of cosmic radiation and atom building had been questioned by a younger generation of scientists led by Nobel Prize laureate Arthur H. Compton. They found it to be inconsistent with the energy generated by the atom building process ascribed to it by Millikan. Rather, the much greater energy released in an "atom annihilation" process appeared to be more consistent with the energy levels required to produce the cosmic ray intensities that were then being observed by an increasing number of researchers.[7]

Although the opposing ideas of Millikan and Compton slowly merged into a state of general agreement over the next few years, much remained to be done to completely understand the origin and composition of cosmic radiation. The "latitude effect," "atomic annihilation," and the modifying effect of the atmosphere along with the creation of cosmic showers were topics that called urgently for further investigation. No one recognized this more than Millikan and, thus, it was no coincidence when, in mid-1933, into this void stepped a brilliant Caltech student named William Pickering who, at that time, was seeking a challenging topic for his Ph.D. thesis in atomic physics.

By the time William Pickering began focusing his attention on cosmic ray research at Caltech in 1933, many improvements had taken place in the experimental techniques employed for their study. Chief among these was the Geiger counter tube, developed by a German experimenter Hans Geiger while working with New Zealand-born Ernest Rutherford at Manchester University in England.[8] The so-called Geiger counter consisted of a short metal tube through the center of which passed a thin tungsten wire that was insulated from the tube by a glass seal at each end. The air within the tube was maintained at low pressure while a high electrical potential was maintained between wire and tube. The passage of a high-energy particle through the wall of the tube ionized the internal gas and resulted in a brief electrical discharge between wire and tube. The resulting pulse of current could be detected by the "kick" of some sort of electrometer instrument. A continuous record could be obtained by recording the electrometer impulses on a suitable strip chart recorder.

When two Geiger counters, one above the other with lead blocks between them, were used to detect the presence of very penetrating particles, two electrometers were used to record side-by-side tracks on a single moving chart. Comparison of the impulses indicated the presence of coincident strikes from a single high-energy particle. In this way, researchers were able to measure the absorption properties of the penetrating particles. A similar arrangement using three Geiger counters could be used to investigate the direction of arrival of the particles. Although these methods represented a considerable improvement over the primitive electroscope device used formerly, they were cumbersome, time consuming, and prone to significant errors.

By 1933 when Pickering entered the field, vacuum tubes had come into use for constructing electrical circuits that would detect coincident discharges from two or more Geiger tubes. Bruno Rossi published a seminal paper in 1933 in which vacuum tube coincidence circuits were used to measure the responses from three counters that were set up in a vertical arrangement to reduce the occurrence of accidental coincidences. His paper attracted a great deal of attention among cosmic ray investigators of the time and demonstrated the advantages of Geiger tubes and coincidence circuits over former methods using the electroscope. It became the way of the future and Pickering decided to start there.[9]

Despite their promise, however, there were inherent problems with the early Geiger tubes and coincidence detectors and Pickering addressed those first. The most important deficiency related to errors in the cumulative counting function and these, in turn, depended upon the stability of the vacuum tube coincidence counters and the stability of the high-voltage potential applied to the Geiger tubes. Pickering reasoned that since the observed coincidences occurred at the rate of a few tens per hour, it would take a very long time to accumulate sufficient counts to reduce the probable error to an acceptably low value. A quick calculation told him that it would take roughly 1,000 hours of continuous recording to accumulate 10,000 counts, the minimum required to achieve a problem error of 1 percent. The long term stability of existing coincidence circuits and high voltage circuits was not good enough to guarantee that level of accuracy in the overall count values, and could well lead to erroneous conclusions in the ultimate evaluation of the experimental results.

When he first started constructing his own Geiger tubes, Pickering followed the example of current investigators. The best of these designs used a central wire of tungsten or copper passing through a brass tube, the interior surface of which was coated with soot. These tubes were unreliable and not uniform in performance and unsatisfactory for lengthy experiments involving coincidence counters. Within a short time Pickering found a better way to build a Geiger counter tube. In his design, a short copper tube, its interior surface made as smooth as possible, supported a central axial tungsten wire by glass seals at each end. After heating to remove traces of grease and create a thin oxide coating, the assembly was washed with dilute nitric and finally rinsed off with distilled water. The counter was immediately evacuated and filled with clean dry air or a mixture of argon and air at low pressure.[10] The completed counters, about 1 inch in diameter and 5.5 inches in length, overcame all of the deficiencies noted in previous designs and proved very satisfactory in the intensive series of laboratory and field experiments to which Pickering eventually subjected them.

By 1933, when Pickering became involved, the problem of reliably and accurately recording the occurrence of cosmic ray coincidences from Geiger tubes had been studied by numerous investigators and several circuits had been noted in the scientific journals. Of these, the preferred method, due to Rossi,[11] used a triode vacuum tube connected to each Geiger tube to detect the passage of an ionizing particle through the tube. The output current from each of the three triodes passed through a common resistance to develop a voltage sufficient to drive an output counter triode into conduction only when all three Geiger tubes were triggered simultaneously. Impulses of current flowing through the counter triode were counted cumulatively by an electro-mechanical counter. Pickering soon saw the deficiencies in this type of circuit and set about correcting them. By substituting thyratron[12] tubes for the triodes in Rossi's circuit, he overcame some of the problems but the current in the thyratrons was difficult to quench rapidly and the counters were expensive to build. Meanwhile, a new type of vacuum tube had come on the market, principally to support the explosive growth in the domestic radio receiver field. It was known as the pentode.[13] Investigators at the Franklin Institute had drawn attention to the possible application of this type of tube to the original Rossi circuit. In pentodes, the anode current cut off very sharply when the grid voltage reached negative four volts, but was restored as soon as the grid voltage returned to its former level. Since Geiger tube impulses always generated much larger pulses than this, even the smallest Geiger impulses could be detected. The impulse current from three or more pentodes connected to a common output resistance generated a voltage pulse of sufficient amplitude to trigger a thyratron connected to an electro-mechanical counter arrangement as shown in Pickering's original circuit diagram.[14,15,16]

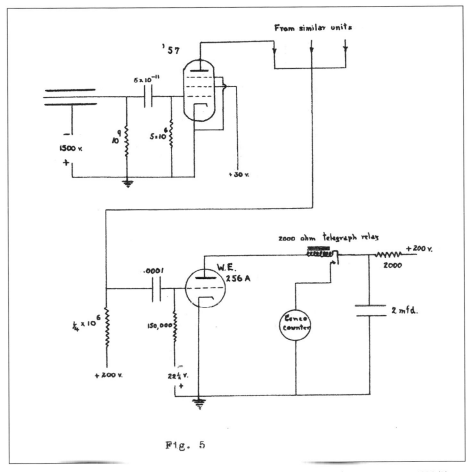

Fig. 5

William Pickering's original thesis; circuit diagram for a coincidence counting instrument for Geiger counters: 1936 (Photo: Courtesy of Pickering Family Trust).

With this arrangement he could count up to 1,000 random impulses per minute, more than sufficient for his experiments. In addition to being reliable, the instrument could be readily adapted to portable form, an important consideration in view of the field experiments that he planned. Although the instrument was further improved in due course (to reduce its resolving time), this was the basic instrument with which he began to accumulate the experimental data that he believed would contribute significantly to our understanding of the origin, composition, and dispersion of cosmic rays. The shadow of the great Millikan hovered over the young scientist as he began to collect his experimental data among the scientific elite at Caltech.

Pickering began by establishing the scope of his investigations. The Geiger counter, he said, naturally lent itself to the investigation of several different types of problems.

There were directional effects in which two or more counters were "arrayed" with their axes parallel, with no material between them. Coincident discharges of the Geiger tubes would then be due to the passage of a single ionizing particle through the set of counters. The direction of incidence would correspond to the solid angle subtended by the array.

If various amounts of dense solid material were placed between the individual tubes in an array, the energy of the rays could be deduced by correlating the observed coincidence rate with the density of the intervening material. When a thin lead plate was placed above three counters arrayed at the corners of a triangle, the observed coincidences would be due to at least three separate ionizing particles emerging, in widely divergent paths, from some point in the intervening lead plate. It would signify the detection of a "cosmic shower." The occurrence of cosmic showers could also be observed in a proximate cloud chamber by using the coincidence signal from the counter to trigger the cloud chamber mechanism.

Finally, there was the controversial "latitude effect," or geographical distribution, of the radiation that could be studied with a portable version of the instrument wherein the Geiger tubes were placed close together to register rays from as large a solid angle as possible.

All of these investigations were possible with his new instrument; with enthusiasm and vigor, Pickering set out to attack them all.

With Geiger tubes suitably arrayed to form a cosmic ray telescope, he investigated the Sun and its surrounding area, and a large region of obscuring matter, several degrees in width, near the constellation of Cygnus. Although the solar experiments were unsuccessful, the observations of Cygnus showed no significant change in the coincidence rate that could be attributed to obscuration of cosmic rays by the dark matter.

Pickering then turned to experiments involving the penetrating part of the radiation.[17]

Previous experimental work by Rossi in 1933 had led to the conclusion that this part of the radiation consisted of high-energy charged particles. With great tact, Pickering pointed out that: "This conclusion is at variance with preconceived notions of the character and properties of such high energy radiation," and that, ". . . it would be of interest to see whether any alternative explanations of this result are tenable."

In a lengthy and complex series of experimental runs involving many different arrangements of coincidence arrays and lead blocks of various thickness, Pickering concluded that: "There is no direct evidence as to the nature of the primary penetrating radiation. However, it can be asserted that this

William Pickering observing solar radiation with his cosmic ray telescope at Caltech, circa 1934 (Photo: Courtesy of Pickering Family Trust).

radiation, for the most part, penetrates lead without appreciable deviation." Occasionally, it generated widely divergent secondary radiation. All of his Geiger counter experiments, he said: ". . . are consistent with the hypothesis that the primary is a very penetrating particle."[18]

The existence of cosmic ray showers had been known for some time and, indeed, had been discovered by experiments using Geiger tubes arranged in a horizontal array. Horizontal coincidences pointed to the occurrence of a group of particles emanating simultaneously from a small region in material above or below the array. The shower patterns were very evident in cloud chamber photographs triggered by Pickering's coincidence instrument, and careful interpretation of these images provided a starting point for his investigations into the shower-producing mechanism. For his experiments he placed plates of lead, iron, or aluminum of various thicknesses above and below two or more horizontally arrayed Geiger tubes. Supported by a theoretical analysis, he evaluated the observed coincidence rates as a function of the experimental parameters; that is, the thickness and type of material. Among the key findings[19] he found that:

a. Shower particles have a penetrating power that, on the average, is less than that required to penetrate 5 cm of lead plate.

b. Showers are caused by *groups of shower-producing photons* rather than single photons that give rise to a series of showers.

c. Showers observed under large thicknesses of lead are caused by photons that are produced in the lead by an (high-energy) incoming ionizing radiation.

Next, Pickering turned to the geographical distribution of the showers. Again exercising his native tact he began:

> It has been known for a number of years that the intensity of cosmic radiation is less in the equatorial band than at other points on

the Earth's surface. This effect has been explained as due to the Earth's magnetic field deflecting incoming particles away from the equator." This raised the question: "Is the shower producing radiation [also] affected by the magnetic field of the Earth?[20]

Pickering set out to find an answer to that question based on the following reasoning:

If the primary is a charged particle with an energy small enough to be deflected by the Earth's magnetic field, then there will be a latitude effect in the number of showers emerging from a given lead plate. If the primary is uncharged, or, if the photon is itself the primary, then there will be no change in the number of showers with latitude.

To investigate the matter he modified his shower coincidence detection equipment to run from a shipboard, direct-current power source and mounted the Geiger tubes on a gimbaled frame to counteract the ship's rolling. Then, with his scientific equipment assembled into a convenient portable arrangement and his new wife Muriel for company, he took a ship from Los Angeles homeward bound for Wellington, New Zealand. During the voyage he would measure the coincidence rate for high-energy particles with the Geiger tubes in vertical array and the shower components with the tubes in triangular array, alternating runs between the two configurations. The ship's log would provide him with latitude and barometric data.[21]

Although a considerable amount of data was collected it was not as consistent, nor as accurate, as he had hoped due to experimental problems related to typical shipboard conditions like vibration, DC power fluctuations, radio interference, etc. Also, being on a north–south course most of the time, the ship's rapidly changing latitude did not allow sufficient time at any given latitude for the accumulation of a large coincidence count. Nevertheless, when the data was evaluated, it did show a latitude effect amounting to about a 15 percent reduction at the equator for the vertical components and about one half of that for the showers. With perhaps an eye to future work, he observed that these experiments had not been extensive enough to get a complete picture of the nature of primary showered producing radiation and suggested that more data were needed, specifically in regard to the effects of high altitude and latitude changes on cosmic ray intensity.[22]

So it was done. Doing the experimental work and writing it up had taken about three years, but at the end it carried him forward to a just reward. Caltech recognized his work as a unique and valuable contribution to the field, and conferred upon him the degree of doctor of philosophy, cum laude. It was indeed a high honor and a great distinction for the young scientist from New Zealand.

California Institute of Technology, Pasadena, circa 1930 (Photo: Courtesy of Pasadena Museum of History Archives, Photo number S-4-22).

During the short time that he spent in New Zealand in 1936 while engaged in the search for the "latitude effect" for his thesis program, he had looked into the possibility of ultimately making a career in the burgeoning hydro-electric power industry. For the past several years, New Zealand had been "electrifying" the entire country, and there appeared to be potential openings for engineers in dam construction, and electric power generation and distribution systems. However, by the time Pickering became interested in job seeking in New Zealand, the worldwide depression had begun to adversely affect the country's economy and he was unable to find the kind of position that was he seeking. Fortunately, before he left Pasadena, Caltech had already offered him a position as an instructor in its department of electrical engineering under Professor Royal Sorenson, and it was to this position that he now decided to return. The couple arrived back in Pasadena at the end of the year (1936) and he took up his teaching duties in the electrical engineering department shortly thereafter. Ten years later, William H. Pickering would become a full professor.

In the late 1930s, the Pickerings moved to a house in the Pasadena city area near the sprawling Caltech campus. At that time the campus included a sizable grove of orange trees where, in later years, the beautiful Athenaeum would be built and where a short 220 kilovolt electrical transmission line was built for

Professor Royal Sorenson to carry out his pioneering research on high-voltage electric power transmission systems. The blend of rural ambience and modern technology created a stimulating environment for the pursuit of science and higher education. As one of Sorenson's young, new instructors, Pickering was required to adhere to the basic electrical engineering curriculum, but his growing interest and, at the time, unique experience with electrical counting circuits persuaded him to introduce these new techniques into the agenda of classical power engineering principles advocated by the head of the department. Within a few years, the principles and techniques that had captured the young scientist's attention as an adjunct to his cosmic ray investigations grew to become an important new field of electrical engineering known as electronics.

Meanwhile, in the loosely knit faculty organization of the time, Pickering continued to work as a valued member of Millikan's cosmic ray research team. He had become acquainted with co-worker Victor Neher in previous years while working on his thesis and, although Neher's special expertise lay with the electroscope experiments, they shared a common interest in Millikan's theories of cosmic ray generation, and their experimental techniques complemented one another in making observations in the field. Neher had worked with Millikan on electroscope observations of cosmic ray intensity at altitudes of 10,000 to 12,000 feet in the Sierra Nevada where temperature control and lightweight packaging were significant factors in the design of the experimental equipment. If Pickering was the coincidence counter and electronics expert, Neher had the experience when it came to lightweight components, temperature control, primary power sources, and packaging.

When Millikan's ideas called for observations at even greater altitudes, beyond the sensible influence of atmosphere, his experimental team of Neher and Pickering turned toward the use of hydrogen filled balloons as the lifting medium and airborne Geiger tubes and instruments with radio transmitters as the observing medium. The new challenge fit perfectly with the complementary skills and common interest of the two scientists. Between them, they quickly developed new Geiger tubes, a suitable high-voltage regulated power supply for the Geiger tubes, the necessary miniaturized coincidence counter, radio transmitter, and antenna—all of it integrated into a light-weight, temperature-controlled flight package called a radio-sonde.[23,24] The balloons and gas filling equipment, VHF receiver, tracking antenna, continuous paper tape recorder, and primary power generator completed the mobile field installation.

After a few test flights to checkout the performance of the airborne package, Pickering and Neher were satisfied that they had the necessary equipment to carry out cosmic ray intensity observations in the stratosphere to altitudes of 80,000 feet and, perhaps, beyond. Moreover, the data could be returned to Earth for recording in "real time." Except for reasons of economical reuse, it was no longer necessary to recover the airborne package to retrieve the data.

Quietly and unobtrusively, a new technology, remote sensing with data retrieval by radio–link, had come into being. Given an appropriate sensing instrument, scientists would now be able to make observations and retrieve their data from places where a human observer could not go. No one, not even Millikan, might have imagined where this idea would lead two decades later.

As Millikan began to embrace the concept of cosmic radiation generation by the annihilation of certain atoms, he reasoned that it should be possible to calculate the energy of the annihilation process from the rest mass of the atoms involved in the annihilation process. Earth's magnetic field, he believed, would act like a huge spectrometer to the incoming cosmic radiation and create line spectra, dependent upon the strength of the magnetic field at different latitudes and the energy of the cosmic radiation. By knowing the cosmic ray intensities at various latitudes, he could calculate the energy required to reach each "latitude of observation" and relate that energy back to its original source, the energy of annihilation.[25]

In addition to this theory, loosely described as the "latitude effect," Millikan had for some time held to a theory that came to be known as the "longitude effect," and along with Neher had published an article on this hitherto unsuspected effect in 1934.[26] The basis for this theory related to the asymmetry of Earth's magnetic polar axis with respect to Earth's geographic polar axis. If this was taken into account, Millikan believed, the spectrographic effects that he sought would also vary with longitude.

To maximize these effects, he planned to make measurements at different latitudes along two widely separated longitudes. For practical reasons, he chose lines of longitude running through the continents of India and North America. He intended to begin with the observations in India and to follow those with observations in North America. It would be a beautiful demonstration of his concept of the theory of particle annihilation.

In 1939, Millikan organized a small expedition to India to demonstrate his theory. Led by Millikan himself, the expedition would include Victor Neher and William Pickering. For Pickering however, the invitation to join the expedition to India created somewhat of a dilemma for, by then, Muriel had given birth to their first child and she was not disposed to remain alone in Pasadena while he traveled overseas on a scientific jaunt of indeterminate duration with the Millikans. It was decided that Muriel and the baby would return to New Zealand to stay in Christchurch with her father-in-law while her husband was away.

However, the following month, shortly after the party reached Sydney, Australia, World War II began with the horrific news of England's declaration of war on Germany. Undismayed, Millikan decided to press on with the original travel plan and sailed for Singapore. Under the wartime exigency, their original ship would take them no further. After a week in Singapore, the party managed to find passage on a dirty old British freighter and, with some trepidation, set sail for Calcutta, India, via Rangoon, Burma. Despite the potential hazard of sailing aboard a British freighter in wartime, the party reached its destination without further incident.

The observations began in the Indian city of Agra. Visits to Peshawar in the northwestern territory and finally down to Bangalore in the southern part of the continent completed the tour. The observations covered about 22 degrees of latitude within a few degrees of a single line of longitude, a sufficient range of latitude to verify the existence of radiation from different atoms being annihilated according to Millikan's theory. Throughout the tour, the Indian government's meteorological service provided the party with supplies of hydrogen for the balloons, recovery services, and weather-related details. Three months later, replete with a surfeit of science data, the party began its return journey to the U.S. by a circuitous route though Hong Kong, Yokohama, and Hawaii.

William left the ship in Hawaii to await the arrival of Muriel on her journey from New Zealand. On their return to Pasadena in mid-1940, the Pickerings resumed the daily routine of life in the academic circles of Caltech faculty and, untouched by the cataclysmic events developing in Britain and Europe, set about raising a child. Pickering meanwhile resumed his teaching duties and began reducing the data from the high altitude flights in India.[27]

Satisfied with the success of the India experiments, Millikan now began planning a similar suite of measurements with his high-altitude balloons on a north-south line across North America. The new measurements would extend from southern Mexico to northern Canada beginning in the Mexican cities of Monterey and Acapulco. Since the latitudes of these two sites fell roughly between those of the northernmost and southernmost sites of the Indian tour, data from them would serve nicely to fill the data gaps in the Indian measurements.

Once again, Millikan invited Pickering and Neher to accompany him and, as before, Mrs. Millikan went along with him. This time though, Pickering also took his wife along on the trip which, they anticipated, would last just a week or two. It all seemed like a great adventure as the expedition set off from Pasadena early in December 1941.

They crossed into Mexico on 5 December and drove south to Monterey, a relatively short journey of about 200 km (120 miles), where they soon found a convenient hotel and suitable site for launching and tracking the balloons. Two days later they launched the first balloon and tracked it for the full duration of its flight. Pickering's radio-sonde worked well and he was able to record good data throughout the entire flight. Well satisfied with their day's work they returned to the hotel to be met by a thoroughly alarmed Muriel who, with her limited ability in Spanish, had just learned from the local Mexican radio that the Japanese had attacked Pearl Harbor in Honolulu. It was 7 December 1941 and now America, too, was at war.

For the second time, Millikan's cosmic ray expedition was caught up, far from home, in the repercussions of a world-wide war. Not without some anxiety, they resumed the balloon flights as planned, collected their data, and apprehensively returned to Pasadena to begin reducing the data and writing the papers.[28,29] They soon found, however, that the local world as they had known it a scant few weeks earlier had changed, and not for the better.

World War II (1942–1944)

Pickering never again took an active part in cosmic ray research. Within a few short months his unique talents were redirected toward the nation's effort to defeat Japan and, in so doing, set him upon a course that would ultimately lead him away from academia and pure scientific research to the cutting edge of applied technology and big science on the grandest scale.

The outbreak of war in Europe had prompted Pickering to file the papers for naturalization as soon as he returned from India. Hitherto, this had not been a matter of any importance in the academic atmosphere in which he moved. But suddenly the world had changed, and the matter of U.S. citizenship took on a measure of considerable significance for him—much greater significance than he could possibly have imagined in the light of what was to happen to the world he moved in, just over a year later. The proceedings were quite straightforward; the U.S. District Court in Los Angeles admitted him to citizenship of United States on 14 February 1941.[30]

As Caltech refocused its powerful research and development capabilities on the war effort, Pasadena developed into a center for industrial research and the manufacturer of precision instruments for scientific research and electronic applications. The national war effort also included the education of vast numbers of servicemen and women in scientific and technical fields beyond the normal high school level and it was here too that the Caltech played a major role.

Very soon after the war started, Caltech moved to a 12-month schedule to support an officer-training program for the U.S. Navy. Named the V-12 program, the courses were designed for civilians from all walks of life who were entering the Navy at the officer level and requiring higher level training in electrical engineering, physics, and mathematics to equip them for further specialized training in Naval schools. As part of Caltech's V-12 program Pickering extended these classes to include electronics, a topic that wartime concerns with radio communications; radar and direction finding, and sonar, had escalated to prime importance. Pickering's demonstrated preeminence in that field placed his services in great demand.[31]

Caltech also became involved in another program to meet the requirements of wartime industry. This was becoming a high-tech war and many of the civilian leaders of industry needed some basic knowledge of science and mathematics at the advanced high school level to properly manage the vast numbers of wartime factories, principally aircraft related, that were springing up all over Southern California. The program was called Engineering, Science, and Management for War Training (ESMWT), and was conducted by special instructors, trained and organized by Caltech, throughout the Los Angeles area. A substantial effort in its own right, the ESMWT was part of a much larger program to harness the scientific talent of the country to support military needs that had been established at the national level by Vanevar Bush, director of the National Defense Research Committee.

It was during the war years, [Pickering recalled] about 1943–1944, that the Japanese bomb-carrying balloons began to appear over the United States. They called them a 'Vengeance Weapon' and had public launchings off the beaches in Japan. They were carrying bombs to the heartlands of America the Japanese propaganda machine proclaimed. It was a clever idea with a balloon designed to reach about 30,000 feet the jet stream would eventually carry it to the United States where instead of dropping sandbags it would drop incendiary bombs to set fire to the forests. A number of these things were recovered between California and Florida because the self-destruct mechanism failed and brought to Caltech [to me] for analysis to find out how it worked because of my association with balloons and electronics. We soon found a fundamental flaw in the design that caused the battery to freeze up and that caused the self-destruct device to fail. One Friday afternoon I was called to the Caltech warehouse where one of these things had been brought in. It was complete, with the self-destruct explosive just sitting there ready to blow-up the warehouse and me with it. I called the Army to come and safe [sic] it before I did any more investigation, but it was a scary experience.

Busy though it was with the normal influx of students and the V-12 and ESMWT programs, Caltech was also involved during the war years with two other programs of great national importance but of quite a different type. One of these was for the U.S. Navy and the other for the U.S. Army. Eventually, Pickering became deeply involved with the latter and it is to this phase of his life that we now turn.

Endnotes

1 Mudgway, Douglas J. Oral history interview with William H. Pickering: Pt. 3. Pasadena, California, January 2003.

2 Goodstein, Judith R. *Millikan's School: A History of the California Institute of Technology* (New York: Norton, 1991).

3 Ibid, Photo: Dedication of 200-inch telescope.

4 Edelsohn, David. "Pre-history of Caltech's South Residence Houses." Pasadena, California: California Institute of Technology; available online at *www.ugcs.caltech.edu: 2004.*

5 Mudgway, Douglas J. Oral history interview with William H. Pickering: Pt. 3. Pasadena, California, January 2003.

6 Kargon, Robert. The Rise of Robert Millikan: Portrait of a Life in American Science (Ithaca: Cornell University Press, 1982).

7 Ibid.

8 Campbell, John. *Rutherford: Scientist Supreme*. Christchurch, New Zealand: American Astronautical Society Publication, 1999.

9 Pickering, William H. "A Geiger Counter Study of the Cosmic Radiation." Pasadena, California: California Institute of Technology, 1936.

10 Gas pressure was 5 cm, equivalent to about 3% atmospheric pressure.

11 Rossi, Nature 125, 1930, p. 636

12 Thyratron was a mercury or argon filled vacuum tube that could be triggered in conduction by applying a voltage pulse of specified amplitude to its grid. Once triggered the grid no controls the current flow through the tube.

13 The pentode had three grids between cathode and anode whereas the triode had only one.

14 Neher, H. V. and W. H. Pickering. "Modified High-Speed Geiger Counter Circuit." *Physical Review*, Vol. 53, 15 February 1938, p. 316.

15 Neher, H. V. and W. H. Pickering. "A Circuit for Rapid Extinction of the Arc in a Thyratron." *Review of Scientific Instruments*, Vol. 9, June 1938, pp. 180—182.

16 Neher, H. V. and W. H. Pickering. "Two Voltage Regulators." *Review of Scientific Instruments*, Vol. 10, February 1939, pp. 53—56.

17 Cosmic rays comprised two classes of radiation, primary and secondary. Primary radiation had great penetrating power and was believed to consist of high-energy particles of extra terrestrial origin. Secondary radiation was shower-like in nature and was thought to be produced as a product of multiple collisions of the primary radiation with atoms in Earth's atmosphere.

18 Pickering, William H. "A Geiger Counter Study of the Cosmic Radiation." Pasadena, California: California Institute of Technology, 1936.

19 Ibid.

20 Ibid.

21 Neher, H. V. and W. H. Pickering. "An Attempt to Measure the Latitude Effect of Extensive cosmic Ray Showers." *Physical Review*, Vol. 58, 15 October 1940, pp. 665—666.

22 Pickering, William H. "A Geiger Counter Study of the Cosmic Radiation." Pasadena, California: California Institute of Technology, 1936.

23 Neher, H. V. and W. H. Pickering. "Light Weight High voltage Supply for Geiger Counters." *Review of Scientific Instruments*, Vol. 12, March 1941, pp. 140—142.

24 Neher, H. V. and W. H. Pickering. "A Cosmic-Ray Radio Sonde." *Review of Scientific Instruments*, Vol. 13, April 1942, pp. 143—147.

25 Millikan, R. A., H. V. Neher, and W. H. Pickering. "Hypothesis as to the Origin of Cosmic rays and its Experimental testing in India and Elsewhere." *Physical Review*, Vol. 61, 1—15 April 1942, pp. 397—407

26 Kargon, Robert. *The Rise of Robert Millikan: Portrait of a Life in American Science* (Ithaca: Cornell University Press, 1982), p.160.

27 Neher, H. V., and W. H. Pickering. "Results of a High Altitude Cosmic Ray Survey Near the Magnetic Equator." *Physical Review*, Vol. 61, 1—15 April 1942, pp. 407—413.

28 Millikan, R. A., H. V. Neher, and W. H. Pickering. "Further tests of the Atom-annihilation Hypothesis as to the Origin of Cosmic Rays." *Physical Review*, Vol. 63, 1—15 April 1943, pp. 233—245.

29 Millikan, R. A., H. V. Neher, and W. H. Pickering. "Origin of Cosmic Rays." *Nature*, Vol. 151, 12 June 1943, pp. 663—664.

30 Mezitt, Beth, Pickering. Private correspondence with the author, August, 2005.

31 Mudgway, Douglas J. Oral history interview with William H. Pickering, Pt. 4. Pasadena, California, January 2003.

Chapter 3

The Cold War Warrior

Jet Propulsion

In the 1930s, the Guggenheim Aeronautical Laboratory at the California Institute of Technology (GALCIT) was one of the world's leading centers for aeronautical research. Nurtured by its brilliant Hungarian-born director Theodore von Kármán, student engineers studied topics in the field of classical aerodynamics such as wing lift and drag, stability of moving bodies, and propeller efficiency. Their thoughts and goals were focused on raising aircraft speeds above 300 miles per hour (mph), improving the safety of air flight, and reducing the costs of air transportation. Wind tunnels were the essential bases for their experiments. In these endeavors they were spectacularly successful, and many of von Kármán's graduate students went on to become giants of the burgeoning American aircraft industry.

A brilliant student among the class of 1936, Frank J. Malina, took a different direction, one that eventually connected with that of another outstanding Caltech physics graduate named William Pickering. Previously, Malina had gone to von Kármán with a far-out proposal for a doctoral thesis. He proposed to investigate the problems of rocket propulsion and the aerodynamic characteristics of atmospheric sounding probes that would be flown to extremely high altitudes by means of rocket motors. Von Kármán was no stranger to the basic ideas of rocket propulsion. In the early 1920s, several years before he came to Caltech, he had listened with interest to the proposals of German experimenters in rocket propulsion, and he was familiar with more recent reports of promising rocket propulsion experiments coming out of Vienna. Von Kármán viewed rocket propulsion as an irresistible new challenge and lost no time in encouraging young Frank Malina to go ahead.[1]

To gain an understanding of the basic principles of rocket motor design and performance, he first built a small rocket motor powered by a mixture of liquid oxygen and methyl-alcohol and set it up in a secure test stand fitted with pressure gauges, flow meters, and thrust measuring devices to measure critical data during each test firing. Obviously such a dangerous experiment could not be conducted in a Caltech with people nearby. After much searching, the group found a suitable site in a dry canyon called Arroyo Seco, a few miles from Pasadena, in the foothills of the San Gabriel Mountains. There, on 31 October 1936, protected by sandbags from a potentially damaging explosion, Frank Malina and his colleagues conducted their first rocket motor tests. Although these tests were primitive when judged by later standards of rocket motor performance, they were a major achievement for the time, and the event and its location became permanently associated with the subsequent history of the area.

By the beginning of 1938 they had accumulated sufficient data for two of them, Frank Malina and Apollo Smith, to present a paper titled, "Analysis of the Sounding Rocket," to a convention of the Institute of Aerospace Science in New York. Later, when the paper was published, it caused a media sensation in the leading papers of New York and Los Angeles where writers fantasized about moon voyages and rocket-powered airplanes.[2]

A few months later, Caltech received an informal visit from General H. H. "Hap" Arnold, then commander of the U.S. Army Air Corps. Always a strong advocate of the application of new scientific discoveries to military purposes, Arnold liked to keep up to date with what was happening in the world of advanced technology. At Caltech, he was particularly attracted by the ideas and demonstrated progress of von Kármán's rocket group. Later, when he met with the Committee for Air Corps Research[3] to recommend topics for future research, he placed particular emphasis on the application of rockets for accelerating the military's take-off aircraft based on the ideas sparked by his visit to Caltech. It was a wild idea, but one that held immense potential for the Air Corps if it could be turned into a practical system. In the end, the committee agreed to provide funding for furthering rocket research at GALCIT with a specific goal of "developing a rocket motor that would be immediately applicable to aeronautical purposes."[4]

Bare necessities: By 1942, GALCIT had constructed some semblance of basic rocket motor test facilities in the Arroyo Seco (Koppes: Wartime foundations) (Photo: NASA/JPL-Caltech Archives, Photo number 1).

Rapid departure: "None of us had ever seen a plane climb at such a steep angle," said von Kármán following the first demonstration of rocket-assisted take-off in America at March Field, California, 12 August 1941 (von Kármán, p. 184) (Photo: NASA/JPL-Caltech Archives, Photo number P381-30).

"Rocket science," no longer a topic of intriguing scientific interest, had become a viable technology whose immediate application was firmly vested in the military.

When the GALCIT rocket group accepted the Air Corps contract it was immediately confronted with the problem of how to make a small, solid fuel motor burn smoothly for 10 to 12 seconds without blowing up. And how could such a motor safely attach to an aircraft to accelerate its take-off run without damage to plane or pilot? Malina turned to von Kármán for help.

Together, they developed a set of mathematical equations that described the physical processes involved in the burning of solid propellants in a constricted combustion chamber. Armed with a better understanding of the interactions of high temperatures, high pressures, and propellant burning areas; and encouraged by additional funding from the Air Corps, the group resumed its experiments in Arroyo Seco. Eventually they found a way to control the rocket burn rate and duration, and verified the design in "static tests" with the aircraft anchored to the runway. They were ready for flight testing.

The first successful rocket-assisted take-off with a small aircraft took place in August 1941 with Lt. Homer Bouchey as the pilot. Less than a year later, the group demonstrated rocket-assisted take-off with a 20,000 pound Douglas A-20 bomber.

In ecstatic words von Kármán saw the event as ". . . the beginning of practical rocketry in the United States."[5] "If we could make a small effective rocket for lifting a plane, then why could we not build a rocket that would lead us into high altitudes and eventually into space?" von Kármán wondered. In those two observations, von Kármán voiced his premonitions for the future with uncanny accuracy.[6]

From about 1940 the activities taking place in and around the shabby buildings set deep in the stony, bleak Arroyo Seco were blanketed in secrecy. No longer a low-key, loosely run field station for a few highly motivated and slightly crazy Caltech students who were testing some far-out science fiction invention; the GALCIT site had become a closely-guarded military facility engaged in work of national importance.

Rocketry became a burgeoning new technology for the nation, and the few nondescript buildings on that improbable site were soon replaced with a great new facility from which sprang the technology that led the nation into space. It would be known as the Jet Propulsion Laboratory. But first, there was a war to be fought and won.

Early in 1943, while the GALCIT rocket group continued its basic research work-up in the Arroyo Seco, the U.S. Army Air Corps began to worry about intelligence reports from Europe that suggested the Germans were developing large rocket-propelled missiles whose range far exceeded anything that was previously known to exist in Germany, or elsewhere. Privy to this information, von Kármán readily acquiesced when the Army Department of Ordnance asked him for a proposal to expand GALCIT's existing rocket engine research program to include the development of long-range rockets.

After some internal exchanges within the Army, the Department of Ordnance accepted the von Kármán proposal with the recommendation that Caltech broaden its research and development program to include not only the rocket motor, but also the development of a prototype guided missile.[7] Although the board of trustees realized that this would carry Caltech into unfamiliar areas of technology, they accepted the challenge and in February 1944 approved a contract with Army Ordnance for a wartime program of guided missile development.

Almost immediately, the GALCIT project was reorganized to accommodate the new change in direction. Additional staff, many from Caltech, was brought in, new facilities were planned, and the facility received a new name. Henceforth it would be known as the Jet Propulsion Laboratory of the Guggenheim Aeronautical Laboratory of the California Institute of Technology, or JPL/GALCIT for short. Operating under its new title, JPL officially began work on guided missile development on 1 July 1944. Owned and managed by Caltech, it had become a contractor for the U.S. Army. However, despite its relationship with the Army and indeed its dependence on the Army for its core programs, JPL retained close ties to Caltech and was able to call upon the Caltech faculty for expert advice and assistance in technical areas beyond its own range of experience.

Spurred by wartime urgency and backed by ample funding from the Army, new laboratories and buildings to house administration and technical staff soon replaced the original dilapidated buildings in Arroyo Seco. A supersonic wind tunnel and several rocket test stands were also added to the facility.

Initially, the new JPL was organized along the lines of its parent organization Caltech, under the chairmanship of von Kármán. However, when von Kármán left Caltech at the end of 1944 to join a scientific advisory board for the Air Force in Washington, the Caltech administration created an executive board to run JPL and appointed Malina to the position of acting director.[8]

Undeterred by the challenge of overtaking the best that Germany could produce, the young men at JPL embarked on a crash program that called for the development of a small, short-range, solid-propellant missile they would call "Private." This would be followed by a heavier longer range version that would include a guidance system and a liquid-propellant engine. That version they would name "Corporal." Later, improved versions would become "Sergeant." In this way the program would progress in stages to reach the ultimate level of performance required by the Ordnance/Caltech (ORD/CIT) contract.[9]

As the work on Private A gathered momentum, it became apparent to Malina and his ORD/CIT team that, impressive though it was, his existing organization was deficient in two areas, both of which would become essential to the successful completion of the Private A program. The areas of concern were test instrumentation and flight testing. His engineers and scientists needed a better way to know what was really happening during engine test firings, for example, and they needed accurate records that could be analyzed after the flight tests were completed. Expert help was needed, and needed fast, to keep up with the frantic pace of the Private program. Malina turned to his mentor, Professor von Kármán.

Von Kármán remembered the smart young physics graduate from New Zealand who had distinguished himself working with Millikan on cosmic ray research just prior to the war. He had joined the faculty in the department of electrical engineering, recalled von Kármán, and was teaching electronics in the V-12 naval officer's training program as part of the Caltech war effort. Von Kármán was aware that, from time to time, Pickering had been called over to Arroyo Seco to bring his experience with electronics and remote control devices to bear on related problems in rocket development. Pickering knew most of the people over there and was highly regarded for his experience, and for his innovative approach to difficult technical problems. Von Kármán sent Malina over to talk to Pickering.

Pickering agreed to join Malina's team at JPL on a part-time basis while he continued his commitments to the V-12 Navy program at Caltech. He was to set up a new section at JPL to provide remote control and telemetry instrumentation support for the ORD/CIT project as a part-time section chief under the direction of Malina. It would be a loose, informal arrangement in which Pickering would remain on the Caltech teaching staff, but would devote part of his time, as required, to working JPL's problems.

It was summer 1944 and von Braun's V-2 missiles, the product of many years of development and testing in the German rocket laboratories led by Wernher von Braun, were raining down on London, England, and, although the world did not yet know it, the war in Europe had just over a year to run to its desperate conclusion.

Before starting in at JPL, Pickering traveled back to the east coast to visit the Massachusetts Institute of Technology (MIT) and the Aberdeen Proving Ground (APG), two major centers of advanced technology related to radar and optical tracking of artillery shells and aircraft, and to look at the state of the art remote control technology at the Sperry Gyroscope Company. He was astonished at what he found there.

As Pickering later explained in a paper presented to the International Academy of Astronautics in 1972: "It was important to recall that the focus of the prewar electronics industry was upon commercial broadcast and communications technology: television and feed back-controlled automation were on the bench, not on the shelf. High-frequency applications—such as radar was to be—were severely limited by the lack of an appropriate amplifying device. It was impossible to buy, and difficult to develop, equipment that would function reliably under the stresses of field operations or rocket flight. But wartime mobilization changed all that.

During this period the state of electronics technology advanced rapidly, almost violently. Anglo-American collaboration made possible a large and growing family of radar equipment and widened fields of application. Components rugged enough to ride an artillery shell, exemplified by the proximity fuse were in production. Aircraft auto pilots, low-noise communications, and fire control systems became widely available. Most of us realized how far the techniques had advanced only when in post-war surveys we observed the extent to which allied efforts had outstripped those of the Germans and Japanese. I later found, for example, that although the V.2 development rounds carried a radio telemetry system, the Peenemünde engineers had to rely principally upon tracking and recovering the wreckage for performance and diagnostic information."[10]

This was the wartime technological background against which Pickering began his association with the Jet Propulsion Laboratory in late 1944 as chief of JPL's remote control section.

Meanwhile, JPL forged ahead with the Private program. Test firings of the Private in December 1944 at a site called Leach Springs, deep in the Mohave Desert had reached an average range of 10.3 miles and provided the Laboratory with its first demonstrated success in rocket flight. However, later tests of a more advanced version of the Private that used wings to increase its lifting capability were largely unsuccessful. They did, however, demonstrate the essential need for a central guidance and control system for the successful flight of a guided missile, a fact that came as no surprise to either von Kármán or to William Pickering.

Early in 1945, about the time the ill-fated Private rockets were being prepared for test firing, the Army approved a JPL proposal to build a scaled-down version of the Corporal that could also be used as a high-altitude sounding rocket. Known as the WAC Corporal, it was to be propelled by a liquid fuel engine manufactured by Aerojet, enhanced by an additional solid fuel booster at launch to give it a launch speed of 400 feet per second. At that speed, the designers believed, gyro-stabilization was unnecessary, since deviation from the vertical would be minimal.[11] Ready to launch, WAC Corporal was 16 feet in height, 1 foot in diameter. The first flights in October 1945 at the newly constructed White Sands Proving Ground in New Mexico proved to be outright successes. Radar tracking showed that it reached an altitude of more than 40 miles, before returning to an impact point alarmingly close to its point of launch. Additional flights were equally successful.

Subsequently, the initial design was improved to reduce weight and to simplify construction, and a five-channel FM/FM telemetry system, based on Pickering designs and built by JPL's remote control section, was added. In addition, the improved version carried a parachute recovery system that could preserve the instrument package on impact, or even preserve the whole missile if necessary, undamaged.

The second set of WAC Corporal tests began in late November 1945 just as Pickering returned from a lengthy tour of post-war Germany and Japan. They took place at White Sands.

Standing tall: Frank Malina with the WAC Corporal in its launch stand at White Sands test range, New Mexico, November 1945 (Koppes: Romantic Rocketry) (Photo: NASA/JPL-Caltech Archives, Photo number P293-364).

For Pickering, these tests formed the beginning of a long and interesting association with the problems of test-range operations and flight instrumentation that would prove invaluable to his career in the years to come.

Speaking of the WAC Corporal he said: "It was a triumphant program in many respects. Outgrowing its research and development function it offered, for the first time, the realistic role of a scientific instrument carrier in a simple, relatively cheap form. It outperformed specifications, exceeding 200,000 ft altitude, a world record at the time. It brought forth a new design cycle for rocket engine and airframe. But most important here, its launch operations, involving the JPL crews, the Aberdeen [radar] tracking team and the White Sands missile range were a valuable preparation for the Corporal [missile] testing to come."

Ultimately, the WAC Corporal became the genesis of the Aerobee—a high altitude sounding rocket built by Aerojet for a program of high altitude research conducted by the Applied Physics Laboratory of Johns Hopkins University, in the 1950s.[12]

A Short, Cold Peace

In May 1945, shortly after the end of the war in Europe, von Kármán led a team of American scientists to Europe on a fact-finding tour of German scientific research facilities specifically related to aeronautics and the development of rocket propelled missiles, particularly the V-2. Von Kármán's group was sponsored by General Hap Arnold, with whom von Kármán had maintained a close working relationship throughout the war. It comprised a number of high-level military personnel that had interests in German military technology, as well as a number of prominent civilian scientists. Among others, von Kármán invited Malina and Pickering to join his group. Hostilities had barely finished when the group reached Paris to begin its tour.[13]

Pickering recollected some of the outstanding events from the tour: "von Kármán was invited up to a place in Denmark to witness the launching of a V-2 that was sponsored by the British group, and I went along with him. One of the constraints was that it [the launching] had to be far enough away so that whatever way it went, it would not get to England. It was a successful launching, and the first time any of us had seen a very large rocket launching. On the way [by aircraft] to the launch site, we stopped at an airfield near Hamburg and as we opened the door of the airplane we were greeted by an Honor guard all drawn up with rifles ready to 'Present Arms.' As we stood in the doorway von Kármán whispered to me, 'What do we do now?' So we walked stiffly down the steps, shook hands with the dignitaries, and we went about our business. We were all honorary Army Colonels and when we had dinner with them that night we learned that they were expecting General Montgomery and his staff, but ours was the first plane to land and so we received the honors."[14]

From Denmark the party visited other parts of Germany to talk to German scientists before moving on to Japan, which had by then surrendered to the allied powers. Von Kármán remained in Paris to complete his report on the group's findings in Europe for General Arnold.

"In Germany," he later reflected, "the aeronautics work was well organized, but the electronics work did not seem to be so well done. The Japanese seemed, for the most part, to be copying what was going on in the West, although they were beginning to develop some ideas of their own." In short, he found very little in Germany or Japan that could add to what JPL was already doing in the field of test instrumentation and electronics development.

In the short, cold peace that followed the conclusion of World War II, William and Muriel Pickering set about readjusting their lives to the new conditions that then surrounded them. By then, William was well on his way to becoming a full professor on the faculty at Caltech and his future in academe seemed assured. Although he could have joined the JPL organization on a full-time basis at any time, he regarded the work there as a research project for the Army that would terminate when the research work was completed and the final reports delivered, and he did not intend to relinquish his chosen career in teaching and research at Caltech for a short-term interest such as that.

Family life for the Pickerings had changed early in the war when son Balfour joined the family in 1939, and daughter Anne Elizabeth arrived in 1943. The family lived in a small house on Craig Avenue in Pasadena during the war. It was a quiet middle-class neighborhood close to Caltech and downtown Pasadena and, when it was time, the children could go to the nearby school. It was a small, but happy and convenient arrangement. However, in 1948, when William's father Bert came over from New Zealand to visit, they decided they needed a larger house to raise their growing children and moved to a beautiful Spanish-style house in Altadena, an upper middle-class area in the lower foothills of the San Gabriel Mountains overlooking Pasadena. The house was large and the children soon made good friends in the neighborhood.[16] There, in 1948, in the quiet, wooded foothill city of Altadena, the Pickerings settled in to raise their family and to establish a professional career in what turned out to be the short-lived peace that followed the end of the "hot war."[17]

Guided Missiles

Things changed at JPL in 1946 when Malina resigned and Caltech appointed another member of its faculty, Louis Dunn, to replace him as director. In keeping with Dunn's personality, maturity, and experience, the organization became more formalized and, as a portent of things to come, JPL's strong relationship with Caltech began to weaken. JPL began to develop a strong sense of autonomy that would color its institutional image for years to come.

The Army exercised minimal supervision of JPL's activities, and the annual review of JPL's progress, and subsequent renewal of the contract, became almost a mere formality. Invariably, the Army approved JPL's funding request for the following year.[18] It followed, of course, that Caltech always received a proportionate share of the contract award to JPL in the form of a management fee that varied from year to year, but was approximately 10 percent. This mutually beneficial arrangement eventually strained the business relationship between Caltech and NASA.[19]

Pickering continued to teach his advanced electrical engineering classes and supervise some graduate students at Caltech until about 1951 or 1952, although this work gradually tapered off as he became more involved with the ORD/CIT contract at JPL. Fully engaged with his advanced engineering classes and graduate students at Caltech, and increasingly involved in the fascinating electronics-related problems at JPL, Pickering paid little attention to the institutional stresses that were developing between the two organizations of which he was an integral part. Confident of a secure future as a tenured faculty member at Caltech, he felt that he could live with the conditions that aggravated so many of his colleagues at both JPL and Caltech.

Despite his part-time status at JPL, he enjoyed considerable authority at the Lab. He hired the best engineering talent he could find to staff the remote control section. He brought in people like Lehan, Cummings, Rechtin, and Parks, former graduate students of his from the early 1940s, to work on the challenging new problems that faced JPL. In the electronics area, Dunn's people did the work at JPL while he provided the direction needed to keep the electronics effort moving forward.[20]

While the main thrust of JPL's effort had been concentrated on Private, planning for the second phase of the ORD/CIT program, the Corporal, had proceeded apace and was well advanced when Pickering returned from his tour with von Kármán.

The Corporal design represented a major technical advance from the Private series in many areas: rocket engine design, aerodynamics and flight path determination, airframe structure, and—most significantly—telemetry and guidance, the two areas for which Pickering was personally responsible, and in which he was rapidly acquiring a prominent national reputation. Telemetry was becoming an essential tool for rocket research and testing.

By that time, Pickering's group had built, and flight-tested, several telemetry systems for use on rocket airfoil test (RAFT) vehicles that had been fired on the Mohave Desert site for aerodynamic research purposes early in 1945. Based on his original cosmic-ray work with the high-altitude balloons, these systems employed three-channel or five-channel FM/FM analog data transmission techniques. They had also been fitted to the later WAC Corporals where they proved very effective in returning the in-flight data. These tests had given the group valuable experience in field operations and telemetry data acquisition. Obviously, these units could easily be adapted for the Corporal flights.[21]

Pickering clearly saw that, to meet its target accuracy goal, Corporal would also have to incorporate what he called a guidance system. Simply shutting down the rocket motor after a fixed interval of time after launch and allowing the missile to coast on a ballistic trajectory to its target, as did the V-2s, would not be good enough for Corporal. With Corporal, its designers were trying to drop a missile into a 1000-ft diameter circle at a distance of 75 miles to 150 miles from its launch point. It was somewhat like trying to thread a needle from a city block away. A specification like that called for a completely new approach to controlling the path of missiles in flight.

As part of his function at JPL, Pickering had been thinking about this problem for some time and had set out his ideas in a technical paper early in 1945.[22]

Pickering's paper was conceptual in nature. It was the task of Malina's engineers at JPL to turn these conceptual ideas into electrical hardware that could be integrated into the overall aerodynamic and propulsion system of the Corporal research test vehicle, and evaluated under real flight conditions at the White Sands test range. Although the initial radio guidance system for Corporal was very rudimentary, it contained all the basic elements required to demonstrate the principles involved in missile guidance. At this stage it was a research vehicle not a weapon system, and understanding the basic principles was the prime objective.

The system envisaged for the first Corporal research test vehicle was similar to that used for aircraft autopilots. Gyroscopes carried within the missile established appropriate flight references for roll, pitch, and yaw. In-flight deviations from these references were corrected by four servo-driven control surfaces, or fins, at the rear of the vehicle. During flight, a ground-based operator kept the missile within the limits of a predetermined trajectory by transmitting to the missile suitable commands for adjusting the reference positions. The operator's job was to guide the missile to the rocket motor cut-off point. After that point the missile would pursue a very precise, free-fall parabolic trajectory to the desired impact point, much like an artillery shell.

Pickering's design included an analog, FM/FM telemetry system to measure the missile response to movements of the control surfaces, fin positions, and aerodynamic loads, and it used the radar beam to not only track the missile, but also to convey the operator-derived control signals to the internal missile guidance system.

Finally, Pickering included a qualifier. He pointed out that this proposal made no provision for controlling the missile velocity. "There is no doubt that," he said, "as shown by the German experience, such control is necessary to attain accuracy without using external control near the target. Accuracy however is not one of the prime objectives of this model, and the complication introduced by velocity control does not warrant an attempt to use it at this time."[23] But accuracy would soon become a prime objective and when it did, Pickering would be ready to deal with it.

Pickering's team at JPL soon turned these concepts into reality. Using special bench-tested components and integrated test assemblies, a complete control system was constructed and tested at the Lab. Of necessity, they used existing gyros and other components that could be adapted for the purpose. In those early days the components required for missiles and rockets were available only "off-the-bench, not off-the-shelf," as Pickering put it. Two years later, the first of the Corporals to carry a simple guidance system, designated the "E" version, was ready for in-flight testing.

Pickering described the first two attempts at guided missile flight in his memoir: "On 22 May 1947, Corporal E, No. 1 rose from White Sands. Weighing almost six tons and stabilized by a pneumatic Sperry autopilot, the slim white rocket lifted off its launch stand, gradually pitched forward toward the target and flew [followed] a ballistic curve to within two miles of its target [at a 62-mile] range.

No precision guidance had been employed, although an experimental radio command was exercised successfully. A ten channel FM-FM telemetry set (actually two of the WAC telemetry sets) returned measurements of guidance and propulsion system parameters, and the trajectory was plotted from radar position and velocity measurements acquired from an active [flight transponder] tracking system."[24]

The JPL launch team at White Sands, led by Dunn and Pickering, celebrated their success with a wild party that night at the military base headquarters. Pickering continued: "Buoyed by this success we prepared the next bird for launching eight weeks later. The second Corporal flight was a fiasco. Apparently the air-pressure regulator which controlled the flow of propellants to the rocket engine malfunctioned; the engine ignited but with insufficient thrust, and burned on the stand for 90 seconds before the rocket was light enough to get off the ground. Then the missile rose, tipped over, headed for the sand and proceeded to skitter through the desert underbrush under power until it blew up. A wag at the scene named it the 'rabbit killer'."[25]

Although the next several flights were only partially successful for various technical reasons, the telemetry system provided excellent records for subsequent engineering analysis of the in-flight problems, many of which were attributed to the damaging effect of the high vibration environment on the mechanically-sensitive electronics components of the guidance system. On the final test in the series, the team was denied even that meager satisfaction when the range safety officer was forced to destroy the missile in flight to prevent it hitting a nearby township. The total failure of Pickering's guidance system shortly after lift-off had allowed the fully powered-up missile to wander, mindless and uncontrollable, beyond the safety limits of the test range.

Somewhat disappointed, but much the wiser for their first experience with missile guidance systems, the team returned to Pasadena to analyze the telemetric, radar, and photographic data and plan the next steps in the Corporal

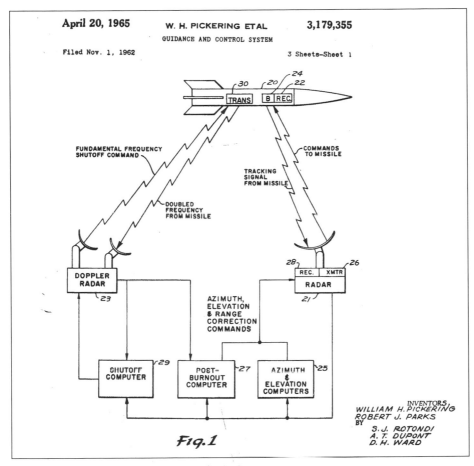

April 20, 1965   W. H. PICKERING ETAL   3,179,355

GUIDANCE AND CONTROL SYSTEM

Filed Nov. 1, 1962   3 Sheets—Sheet 1

Fig.1

INVENTORS,
WILLIAM H. PICKERING
ROBERT J. PARKS
BY
S. J. ROTONDI
A. T. DUPONT
D. H. WARD

The basic concept of a guidance and control system for the Corporal missile due to William H. Pickering and Robert J. Parks. It employed an SCR-584 radar beam-riding guidance system with a separate Doppler velocity measurement system to cut off the rocket motor at a prescribed point on the trajectory. United States Patent 3,179,355 was issued to Pickering and Parks as Inventors in April 1965 (William H. Pickering: personal papers) (Photo: Courtesy of Pickering Family Trust).

guidance program. For the next couple of years the engineers at JPL redesigned, built, and tested various elements of the guidance system including the ground radar to improve its in-flight performance, resistance to vibration, and reliability.

During these initial flights, Pickering had been very much aware of the shortcomings of his original design. As he pointed out in his proposal, it made no provision for measuring the missile velocity, a critical factor in the ultimate accuracy of the overall guidance system. Working on this problem with Robert Parks, a former graduate student that he had hired into JPL, he had developed an idea to correct that deficiency by making use of the "Doppler effect," or change in frequency of a radio signal transmitted from the missile as it receded rapidly from the launch site after firing. A special Doppler radar would be used to send a precisely known radio signal up to the "bird" during flight. The bird would retransmit the uplinked signal back to the ground where the radar receiver would compare the received signal with the transmitted signal to extract the difference in frequency due to the receding motion of the bird. The so-called Doppler frequency would give him a direct measure of the velocity of the "bird." The basic idea is illustrated in the diagram

In the improved version of the Corporal guidance system, this critical "velocity" parameter would be provided by the Doppler radar.[26]

While his group at JPL worked on improving the Corporal guidance system, Pickering continued his academic work at Caltech and also devoted some of his attention to the matter of optical and electronic instrumentation facilities for flight testing long-range missiles. Earlier, he had played a major role in setting up the electronic instrumentation, radar, and telemetry for the test range at White Sands. In fact, JPL's WAC Corporal firings had inaugurated the new facility as its first paying customer back in 1945. He had accumulated a great deal of experience in the instrumentation and operational complexities of rocket testing. When the U.S. Navy began to set up a Naval Air Missile Test Center (NAMTC) at Point Mugu, near Oxnard, California, it turned to JPL for advice and consultation and Pickering directed several studies to help the Navy set up an instrumentation system to suit its purpose there.[27]

As if this was not enough, Pickering also served as chairman on a committee sponsored by the Research and Development Board of the Department of Defense (DOD) to establish and coordinate standards for telemetering systems the military test ranges then being constructed throughout the country. At the time, Pickering's analog FM/FM telemetry system, which had a proven record of performance, was selected as an interim standard pending the development of a pulse, or digital, system. As the military needs for in-flight missile testing expanded, test range instrumentation became complex and the committee assumed a broader role as the Inter Range Instrumentation Group (IRIG). The IRIG standards, adjusted to accommodate new developments in digital techniques, formed the basis for all telemetry systems for DOD missile ranges in the U.S. far into the future.[28]

By 1949, William Pickering, then age 39, had become a very busy young man. Engaged with Corporal guidance problems at JPL, his graduate students at Caltech, instrumentation systems for military test ranges and JPL's wind tunnels, and driven by a persistent interest in the distant potential of rockets for high-altitude research; he nevertheless enjoyed watching the progress of his rapidly maturing family. His widowed father, Bert, had by then returned to New Zealand to live out the remainder of his life with cousins in Auckland.

His children were growing up fast and doing well at school and, as time permitted, Muriel began to develop associations with women's groups at Caltech and Throop Memorial Universalist Church and to become involved with the local community affairs, particularly those relating to the local library.

As the communist threat to America's security began to loom larger on the international horizon, Army Ordnance began to focus more intently on the results of its rocket research program at JPL. In the five years since its inception, the Corporal guided-missile research project had shown great promise. Four test flights had been carried out, one of them highly successful, and valuable performance data had been obtained from the others. JPL believed that it now understood the fundamental problems of propulsion and guidance as they related to guided-missile technology. In mid-1949, with all of this in mind, Colonel (later Major-General) Toftoy of the Missiles and Rocket Branch, U.S. Army Department of Ordnance, approached the director of JPL with a serious and far-reaching question: "Would it be feasible," he asked Louis Dunn, "to convert the research-oriented Corporal test-vehicle into a military weapon system capable of carrying a warhead with great accuracy to a specified target, and capable of operation under field-combat conditions by suitably-trained service personnel?"

It was a question that provided much food for thought indeed, and Dunn took it upon himself to provide an answer. He called on his colleague Bill Pickering for help.

With typical brevity of statement, Pickering recalled the decision-making process, "Dunn brought me into the inquiry and we gave the matter much thought before deciding in effect, 'Let's give it a go'."[29]

In September 1949, Dunn and Pickering made the long train trip to Washington, DC, to confer with Colonel Toftoy at the Pentagon. It was a very unpretentious meeting, just four of them, two military officers and two scientists around a table in a small Pentagon office. Toftoy described the Army's need for a demonstrably field-worthy guided missile, while Dunn and Pickering tried to explain what they thought they would have to do with the existing Corporal design to make it so. Both men were impressed by Toftoy's confidence in JPL's ability to undertake what was essentially an industrial-development task considering that JPL was a research-oriented organization and had no experience in that field. "We were being pretty naïve about it [the complexities]," Pickering recalled, "but they [the Army] didn't really understand the complications either. They were sufficiently impressed by our record that they believed it would go all right. I believed we could handle the guidance problems and Louis was confident that we could handle the industrial-engineering transition. It really was as simple as that; we said we could do it, and Toftoy said, 'Go ahead'."[30]

Led by Louis Dunn and William Pickering, JPL was about to become involved in its first guided weapon development program for the U.S. Army.

When they returned to Pasadena, Dunn asked Pickering to phase out his academic work at Caltech and to take charge of the Corporal weapon development as a full-time project manager. Pickering agreed, and few months later, the program was underway.

Pickering's moral view of the work that he was undertaking, at the time, is reflected in the following conversation with the author:

> DJM: Did you (as a professed man of science) have any moral scruples about the end use of the wonderful technical device (Corporal) that you were developing?
>
> WHP: Oh yes, I did, from time to time, because we knew that the Corporal as a weapon was being thought of by the Army as a means of carrying an atom bomb. The whole idea of that was an unpleasant thing to think about in detail for me. On the other hand, the satisfaction of being able to oversee the technical design and solve the technical problems that enabled you to throw a missile on a target one hundred miles away was pretty challenging and very satisfying when you showed you could do it.
>
> But I thought of myself as contributing in a minor way as far as the total picture was concerned. We were not making the ICBMs (intercontinental ballistic missiles) that were aimed at destroying cities. We were building a device that would be a superior weapon on a battlefield.
>
> DJM: You perceived that the technical challenge was sufficient reason to be involved and you did not concern yourself about the ultimate end result. There were others whose job it was to worry about that.
>
> WHP: That's true. The people who were doing the planning and application to particular situations were the one who should worry about that.
>
> You have to put the whole thing against the background of Cold War development. We knew that the Soviets were developing an ICBM, and you might ask did the personnel at the Lab as a whole worry about these issues and was there any discussion about moral problems? I do not remember any such discussions. I think that they simply accepted it as an interesting technical job, with a lot of interesting features, and got on with it.[31]

When Pickering began work on the "weaponization" of the Corporal, the propulsion and aerodynamics systems were quite advanced. Both he and Dunn believed that all that was needed in those areas was refinement, or optimization, of already demonstrated systems and hardware. These were largely matters of weight, material, and dimension; although the most critical technical problem, that of long-range accuracy, remained a significant problem in Pickering's mind. Subsequent events, however, were to prove them wrong.

Problems in manufacturing components for the Corporal missiles to meet the required standards of reliability proved to be more complex than they anticipated and forced the introduction of strict new controls into the manufacturing processes. In due course, the quality of the completed missiles improved to the point where, by August 1952, Pickering was ready to give approval the start of test firings of the contractor-built Corporals at the White Sands test range.

Pickering had been associated with the White Sands missile range since the early WAC Corporal firings in 1945. Although it was essentially a military establishment, he and his JPL team of civilian scientists and engineers were highly regarded, not only for their highly-advanced technical expertise but also for their very "laid-back" social attributes. Under the direction of their leader Bill Pickering, they manifested their awesome technical prowess on the test range by day, and their equally awesome social attributes on the military base and in the nearby towns of Las Cruces and El Paso by night. Also their presence was frequently in evidence between the U.S. and Mexico, in the Mexican border town of Juarez.

Pickering was no stranger to field testing. He believed strongly in its worth, and he had an imposing record of cosmic ray work with Millikan, under extreme duress at times, to show for it. All engineers, he insisted, should be prepared to demonstrate, personally, the efficacy of their designs in the field. It was not a responsibility that could be delegated. Besides, the shared experience of intense and at times stressful experience of away-from-home test trials forged a common bond of dependence and trust and confidence among the members of the technical teams that were involved. The abiding nature of these relationships would carry many of the White Sands test teams far into the future fortunes of JPL.

Pickering enjoyed recounting a story related to this period of his life that concerned the introduction of air brakes for missile aerodynamic control

> It was easy to accelerate a rocket but how do you slow it down? At JPL, we had been debating that question for some time. One night, four of us were driving back from Las Cruces after dinner where we had been discussing that problem when suddenly, one of the aerodynamic engineers in the car said he had a great idea that he would demonstrate to us. When the car came to the top of the next hill, he told the driver to switch off the engine and coast full speed down the hill. When the car was really going fast, two of us pushed the two rear doors open against the rushing wind, and the car almost came to a stop. It was a practical and powerful demonstration of the efficacy of drag brakes at high speed that we used in later designs for Sergeant.[32]

It was against this background that JPL began test firing the first of the contractor-built Corporals in August 1952. It was Pickering's project and he led the way. With Parks and Rechtin managing those sections related to the guidance and telemetry, areas for which his JPL division was responsible, he was able to take a broad overall interest in the missile and its performance. And what he found was not very good.

JPL's highly skilled and experienced crews fired 56 production missiles in that first year of test firings at White Sands. However, only 43 percent were considered successful, a value that fell far short of the 95 percent success rate required by the military specifications. Furthermore, production capacity had fallen far short of the goal of 20 missiles per month desired by the Army. A mere 13 missiles was the best run the contractor ever produced in any single month.

Matters became further complicated when JPL began training Army troops to prepare and launch the missiles. This training task went far beyond anything the lab had experienced before. It was even more foreign to JPL expertise than industrial production, and JPL did not do well. Dunn and Pickering were becoming engaged in a task that, essential though it was, held no challenge, interest, or attraction for either of them.

By mid-1954, the project had become a mess and an impasse loomed ahead. To head off a crisis and attempt to improve working relationships and communications between the three organizations, the Army appointed a technical committee led by JPL to coordinate the overall activities and resolve mutual problems expeditiously.

The work went ahead and eventually troop training improved, but the success rate of Army-directed test firings at White Sands rate fell far short of Army expectations, and the usefulness of the Corporal system as an effective military weapon came under serious question. As Pickering and Dunn had made clear at the outset, the missile contained many other shortcomings that were a consequence of its nonmilitary origins and such problems could not be corrected by any amount of redesign.

The Army, however, had no option. Pressured by the tense international situation in Korea and Europe that engulfed the nation in the mid-1950s, it went ahead and, a year later, deployed the Corporal as a surface-to-surface guided ballistic missile in Europe. Thus, Pickering's Corporal project produced the first guided missile of that type to see operational use for the U.S. in the Cold War era. Despite its shortcomings, it was an awesome weapon when tipped with an atomic warhead, although it may well have been more effective for its psychological value as a deterrent rather than its tactical value as a weapon.

Fully engaged with the exigencies of the Corporal program in mid-1954, Pickering was taken completely by surprise when, quite suddenly, Louis Dunn resigned from JPL to join a newly-formed missile engineering company in Los Angeles that was engaged in building Atlas, the United States' first intercontinental ballistic missile. When Caltech president Lee DuBridge asked Dunn to suggest a replacement for his position, Dunn, equally unexpectedly, nominated Pickering. Although DuBridge knew that Pickering's field of expertise lay in missile guidance, telemetry, and electronics, Dunn convinced him that Pickering was at once "the right man in the right place at the right time." President DuBridge called Pickering over to the campus for a chat.

Recalling that afternoon meeting many years later, Pickering said:

> I was surprised when he called me, quite frankly, because my thinking of the Laboratory was that it was still primarily aeronautics and chemistry and mechanical engineering with electronics just tacked on. And so the idea that I, as an 'electronicker,' would be offered the job did not occur to me. But, I had no hesitancy in deciding [to accept]. I had been with the Lab long enough to know what was going on and how thing were done up there. I was definitely very happy to take it on. DuBridge asked me if I would treat it as a stepping stone to some big industrial job in the sense that Louis Dunn did, and I assured him that I would not, and that I wanted to come back and resume my job as a Professor. He said, 'OK I will give you leave of absence'.

Wanting to "come back as a professor" was a reflection of Pickering's view of the JPL situation at the time. He knew he would not be happy just making a succession of weapons for Army Ordnance. Up at JPL they were working on the Corporal, and studies for the follow-on Sergeant were well advanced. He felt obligated to see that work completed, but did not want to do another one [weapon]. "It was a sort of qualified answer," he said, "[Rather like] I would take it for the time being." DuBridge agreed, and gave him leave of absence from Caltech—for more than 20 years as it turned out—but no one foresaw that happening then.

In pensive mood, Pickering drove back home to Altadena for a quiet celebration with his family. Unaware of the onerous responsibility that he had just assumed, the family knew only that he had "got a big new job" and they were happy for him.

Everything changed for Pickering after that. No longer a college professor working in a narrow field of interest of his own choosing, he now had to step back and take a much broader view of the Laboratory's activities. By that time the population of the lab exceeded 1,000, and its annual budget was more than 11 million dollars. With its Army contracts as the dominant activity JPL had become a significant organization in its own right. Robert Parks took over Pickering's position as project manager for Corporal, while Pickering addressed organizational matters and began to make his presence felt as an administrator.

Koppes aptly described the new regime: "More personable and less of a martinet than Dunn, Pickering was also less rigorous as an administrator. . . . Some interpreted Pickering's laissez-faire attitude toward management as indecisiveness. . . ." Nevertheless, the transition to the new director was smooth, relations with the Army and the Corporal contractors were improving and

JPL Director William Pickering at the firing console for the launch of Corporal round 100, at White Sands missile range, April 1955 (Photo: NASA/JPL-Caltech Archives, Photo number P19445B).

just ahead lay a tantalizing new challenge for JPL, a follow-on contract for the guided missile weapon system, Sergeant.

While Pickering settled into the director's office at JPL, test firings of the Corporal continued at an ever-increasing pace at White Sands. When the test program was finished in 1955, they had launched over 100 Corporal missiles.

Although he was no longer directly involved in the test firings, Pickering exercised his prerogative as director by electing to press the firing button for Corporal round 100. Corporal round 100 was set up on its launcher and programmed for an impact point due north of the launch area. The weather was perfect, photographers were on hand, and the countdown proceeded smoothly to the last few tense seconds. Five, four, three, two, one: FIRE. Confidently, the new director pressed the red button on the firing panel. The 40-ft long, slender, white missile rose majestically straight toward the heavens and began a graceful turn towards . . . not the north as programmed but, to the east, and ignominious destruction by the range safety officer. Said Pickering with a twinkle in his eye, "They never let me fire a Corporal again, even if I was the Director."[33]

Shortly before Pickering took office, a special Ordnance-based committee evaluated proposals from three study teams, including JPL, for a follow-on to the Corporal. Based on the committee's recommendation, the Army accepted JPL's proposal and issued a new contract to that effect in late 1954. The new weapon system would use a solid-fuel propulsion system to avoid the cum-

Sergeant, an inertially guided, solid-propellant, surface-to-surface, ballistic missile launches from White Sands missile range in the course of JPL's four-year development and production program for the U.S. Army, 1956-1960 (Photo: NASA/JPL-Caltech Archives, Photo number 293-3302B).

bersome ground support systems required by Corporal, it would use an improved radio–inertial guidance system to avoid interference from enemy counter-measures, and range and accuracy would be improved. It would be designed, from the start, as a complete weapon system. The Sergeant weapons system, including troops trained in its field operation, was to be ready for service by 1960.[34]

Determined to pass on all of the experience they had accumulated on the Corporal program, Pickering appointed Robert J. Parks, long time colleague and chief of the JPL's guidance research division, as project manager for Sergeant. Under Parks's direction, the ground-to-air radio links were simplified, transistorized electronics and new packaging was implemented to reduce vibration, and automatic checkout equipment sped up launch readiness procedures. Fitted with aerodynamic drag brakes to reduce the missile velocity to the desired value, and inertial guidance system, Sergeant met or exceeded its targeting accuracy goals and provided a high measure of freedom from enemy radio interference.[35]

Said Pickering:

> The Sergeant missile development went forward at a more rapid pace than Corporal, partly because of the simplicity and excellent performance of its power plant, but mostly because both engineers and soldiers were building on the Corporal experience. Less than three-dozen Sergeant missiles were test flown at White Sands, compared to over one hundred Corporals. . . . Corporal had taught us how best to use industrial collaboration. The engineering firm that would build the Sergeant, a new division of the old Sperry Gyroscope Company that I had visited on my first trip out of JPL a decade before, joined [JPL] as cocontractor during Sergeant's development.[36]

Speaking of the apparent contradiction between his Laboratory's engagement with missile development and its primary mission of applied science, Pickering observed: "the ultimate positive effect of missile development on applied science was almost a case of serendipity rather than technological determinism—the missile contributed much of the technology for, as well as the impulse toward, the practical science of astronautics."

When he took over as director these intertwining streams of activity, missile development, and applied technology coalesced into a broad stream of endeavor that began to carry the Laboratory forward, with increasing speed, toward space.

Toward Space

Pickering's interest in upper-atmosphere research went back a long way.[37] Ever since 1946, when he returned from the post-war tour of Germany with von Kármán, he had been involved in the ultimate destiny of the V-2 rockets that had been brought to the U.S. from Europe, and the German "rocket scientists" that accompanied them. In this country, the V-2 program was directed by the Army Ballistic Missile Agency (ABMA) and managed by the Army's Redstone Arsenal in Huntsville, Alabama. When Army personnel, assisted by the German engineers then assigned to the staff at Huntsville, began test-firing the V-2's at White Sands late in 1946, a group of American scientists, including Pickering, were quick to seize the opportunity to substitute a science payload for the missile's warhead to gain reliable and virtually cost-free access to the upper atmosphere and the regions beyond. The Pickering-designed telemetry system was easily adapted to provide a data return channel for the upper atmosphere science measurements. Working with this program at White Sands, Pickering inevitably made the acquaintance of the German experts including von Braun and James Van Allen, an eminent scientist from the University of Iowa. So began a lifelong association that would bring worldwide, public acclaim to all three men in the years ahead. A new era in the field of upper-atmosphere research was about to open and Pickering, Van Allen, and von Braun were at the forefront.

Although rockets appeared to be attractive vehicles for high altitude research, they had a serious deficiency for cosmic ray investigations due to the short duration of their flight above the atmosphere. Pickering recognized the problem in a 1947 paper when he suggested that: ". . . extensive cosmic-radiation studies be deferred until a satellite rocket can be produced."[38] He did not predict when that might become possible. Nevertheless, he sustained his interest in the subject and became a strong advocate for a rocket-propelled scientific satellite when the necessary technology became available.

When the National Academy of Sciences established a committee to consider options for the United States' participation in the International

Geophysical Year (IGY),[39] Pickering joined a subcommittee of the rocketry panel. It became known as the long playing rocket study group, so named after the latest device for playing long-duration audio recordings that had just come on the market. In this group, he and others "reported on the usefulness of a satellite in the IGY program without advocating any specific project." In October 1955, when the IGY committee organized a technical panel for the Earth satellite program, Pickering became a member and, throughout the life of the IGY, headed a working group on tracking and computation of satellite orbits.

Project Orbiter

In mid-1954 Pickering took part in a symposium to discuss the potential for building and launching an Earth satellite as part of the United States' contribution to the IGY. Engineers and scientists from the Office of Naval Research, the American Rocket Society, the astrophysics community, and the Army Ordnance Rocket Program, including von Braun, attended. Discussion was vigorous and there was no lack of ideas. Influenced no doubt by Pickering and von Braun, the outcome of the proceedings led to a proposal to put a lightweight, inactive satellite into Earth orbit using a Redstone missile as a first stage, with a cluster of JPL-developed solid rockets as the upper stage.[40] It was an innovative idea that made use of existing technology from ABMA and JPL to create an entirely new device that pushed upper atmosphere science beyond anything formerly possible. The proposal was called Project Orbiter. Calculations showed that the idea was entirely feasible and, for the next year, Project Orbiter was the subject of much study and debate. After considering its options, JPL decided that its well-tested 1/5 scale motors from the Sergeant program would be most suitable for the Orbiter upper stages. Studies showed that, using the Redstone and scaled-down Sergeant motors, both flight-proven hardware then available, the Project Orbiter in this form could be launched by August 1957, just a few weeks after the planned opening of the IGY observations period. It was submitted to the IGY committee for consideration as a joint ABMA-JPL proposal.

Along with the ABMA-JPL proposal, the IGY groups, including Pickering's technical panel, evaluated a competing proposal called Project Vanguard that had come from the Naval Research Laboratory. Unlike the Orbiter, Vanguard used a non-military rocket to carry a small science package into Earth orbit and was provided with a radio tracking system. Although the Vanguard rocket had yet to be proven in test firings, its nonmilitary origin was perceived as a favorable feature in the context of the times, and the science package and radio tracking system were seen as additional advantages over the Orbiter proposal. Realizing the comparative shortcomings of the Orbiter proposal, Pickering's team scrambled to adapt their existing telemetry and radio guidance technology for use with Orbiter. Although this effort produced an effective, ground-based tracking system known at JPL as Microlock, the effort was to no avail. President Eisenhower had previously expressed a strong desire to keep the nation's scientific satellite experiments out of the military domain, and the DOD committee responsible for making the ultimate decision ruled in favor of the Vanguard proposal for America's first Earth satellite program.

The Orbiter proposal was shelved and Pickering, despite his disappointment at the committee's decision, turned his attention to ". . . organizing the operational support for the Vanguard mission. It was part of our (the IGY working group) task to encourage and coordinate various efforts, including international activities, in connection with the observation and tracking stations for Vanguard."[41]

Project Vanguard

As part of the United States' contribution to the IGY, Project Vanguard was not encumbered by the security constraints that covered all of Pickering's work for the military establishment. Pickering was quick to take advantage of that fact to promote his ideas for Earth satellites. With this as his principal topic, his remarkable propensity for public speaking soon manifested itself in an ever-increasing number of speeches and lectures in the public sector. It began in Los Angeles in October 1956 with an address on Project Vanguard to a group from the motion picture industry. In the light of events that would follow just one year later, this speech became a landmark event in the sense that it marked Pickering's first public statement of the United States' intent, and capability, to put a satellite into orbit about Earth for purely scientific purposes. His speech was titled "Project Vanguard: The Earth Satellite Program."[42]

He began with a description of the IGY program that confirmed the United States' intent to launch an Earth satellite the following year.

> The earth satellite program of the United States is part of the International Geophysical Year Program. During the period from July, 1957 to December, 1958, there will be a concerted effort on the part of most of the nations of the earth to gather data of geophysical interest. . . . The international committee which is coordinating the IGY program . . . recommended in October of 1954, that a satellite program would be of extreme value as part of the IGY activity. In July of last year, the President announced that the United States would indeed fly a satellite as part of our contribution to the IGY program. Since then, the Russians have announced that they also will fly a satellite. The President has stated that each of the services will contribute to the effort, but the primary responsibility has been given to the Navy. The Naval Research Laboratory has set up what is now known as Project Vanguard.[43]

In his typical tutorial style, Pickering continued with a most lucid, but quite technical, description of the physical principles underlying Earth-orbital flight, the rationale for firing the booster rocket in an easterly direction to take advantage of the rotation of Earth, and the way in which small variations in the satellite's launching direction affect the shape of its ultimate orbit, and its inclination to Earth's equator. "So to sum up," he said, "we must take our satellite object to an altitude of 300 miles and launch it with a velocity of five miles per second in a specific direction with an accuracy of about one degree." To his audience, Pickering might have conveyed the impression that "that was all there was to it. Putting a satellite in orbit was quite straight forward, just a matter of getting the numbers right." But no one knew better than Pickering that was not the case. It was an incredibly difficult matter, as events would soon show.

Concluding, he said, "The earth satellite program has stirred the imagination of scientists and the public alike. It will be a spectacular demonstration of the potentialities of modern technology and the first real step towards the conquest of space."

It was a masterful presentation that covered all aspects of satellite flight, at a level that did not condescend to the understanding of his listeners and yet held their interest and rewarded them with a fascinating view of the future of space-flight. His ability to "hit the right level" and his articulate and compelling delivery would become the hallmark of Pickering's public speeches from this point on.

In February the following year (1957), he presented a seminar on "Some Problems Associated with a Small Earth Satellite" to graduate students at Caltech.[44] Relaxed, and obviously enjoying his return to campus as a lecturer, he used only a few handwritten notes to cover the complex mathematical principles involved and the practical engineering specifics required to realize the world's first spaceflight.

A few weeks later he addressed the graduating students of a small technical university in Los Angeles on the topic of "The Engineer and the Next Ten Years."[45] In this speech he emphasized vistas of endeavor and moral obligation that reached beyond the more traditional confines and physical limitations of the engineering profession.

As his public appearances began to attract attention in the area, an increasing number of invitations and requests for presentations to professional societies and educational institutions began to arrive at the JPL director's office.

In April 1957 he presented the annual Faraday lecture to the graduating students of a local high school.[46] For this speech he returned to the topic Project Vanguard with several important additions, which he went to considerable pains to explain. These represented the products of his IGY working group on tracking and orbit computation. They involved the "Moonwatch" network of optical telescopes to be established by the Smithsonian Astrophysical Observatory, the "Minitrack" network of radio tracking stations being set up by the Naval

Research Laboratory, and the informal amateur radio community that was being encouraged to listen for the satellite signal and send in reports. Between them, these groups formed a dense, worldwide set of observing stations whose reports would be used to calculate a very precise orbit for Vanguard.

After again describing the inherent scientific knowledge that could be deduced from a precisely-known orbit, Pickering speculated on the future of space research.

He said:

> Before the end of the IGY, I believe we will see both United States and Russian satellites flying around the Earth, collecting data from outer space, sending us information on the conditions they observe, and providing us with a new type of astronomical object and a new geodetic tool for measuring the Earth. It might even be possible to send up some simple television devices and look at the heavens from a completely new viewpoint. Perhaps the astronomer's dream of a telescope outside the Earth's atmosphere can be fulfilled. It is quite certain that bigger and better satellites will follow the simple IGY devices. Probably both Russia and the United States, and possibly some other countries will build them. When we succeed with satellites of this size it will be relatively easy to send a smaller object on a longer trip through space—perhaps to circle the Moon.

In the light of later events it was a remarkably prescient view of the future, emphasizing the immediate utilization of space for scientific purposes and suggesting the further possibility of manned spaceflight.

Traveling Man

Bill Pickering had always been an avid traveler. He loved to see new places, meet new people; he was always "on the go," making "deals," and engaging in arguments where he advocated ideas for an Earth satellite program, or promoted JPL's interests and protected his control over it. The constant flux and pressures of life on the move seemed to provide a stimulus for his enthusiasm and an outlet for his inexhaustible energy. Elevation to the position of director increased, rather than diminished, his need to travel. Pasadena to White Sands, Huntsville, and Washington became a regular commute. He recollected "red-eye flights out of Los Angeles at 11:00 p.m. to Washington, DC, a day full of meetings and presentations at the Pentagon and a 5:00 p.m. return flight to Los Angeles. Muriel did not complain, but she clearly was not happy about it."[47]

It was all classified work and his wife could not know what he was doing. She perceived that he was away from home most of the time and accepted that it was not conducive to any significant social life. She graciously managed the major tasks of dealing with the needs of the rapidly growing children, and she sought avenues of interest to satisfy her own intellectual talents. In later years, Beth Pickering Mezitt remembered her mother's public and private interests:

> Not only was she active in the League of Women Voters and the PTA, but she was also a 'Friend' of the Altadena Public Library, and a founding member of La Cañada Valley Beautiful," she said. "We faithfully attended the Throop Memorial Universalist Church in Pasadena. Many of the members were from Caltech and it was considered rather an intellectual approach to religion. But, I think my Dad's religion was internal, and not dependent on the formality of organized religion. Certainly his sentiments for the stars and solar system and the wonders of science, and how it all works, can only be classified as reverence.[48]

Bill Pickering liked the outdoors and rediscovered an extension of his Havelock boyhood in a love of fishing—especially fly-fishing. He would often take his family on long hikes in the local mountains and on other occasions for longer vacations in the Sierras where he could practice his fly-fishing skills to good effect.

It was a world apart from the deadly work in which he was engaged and could share with no one, not even Muriel. They could not know it then, but all that was about to change.

Endnotes

1 Koppes, Clayton R. *JPL and the American Space Program: A History of the Jet Propulsion Laboratory* (New Haven: Yale University Press, 1982).

2 Ibid.

3 The Committee for Air Corps Research was one of many specialist groups under the National Academy of Science. Both Millikan and von Kármán were members, a fact that, no doubt, had considerable bearing on its support for Arnold's recommendation.

4 Koppes, Clayton R. *JPL and the American Space Program.*

5 von Kármán T. *Pioneer in Aviation and Pathfinder in Space* (Boston and Toronto: Little, Brown and Company, 1967).

6 In the aftermath of this event, von Kármán and others from the group founded a small commercial company to manufacture the Jet Assisted Take-Off (JATO) units in large quantities for military purposes. They named it the Aerojet Engineering Company. The company flourished and eventually became a major U.S. corporation for the manufacturing of rocket engines of all types.

7 Koppes, Clayton R. *JPL and the American Space Program.*

8 Ibid.

9 The successive stages of the program were named for Army ranks to preserve their association with the Army and to connote increasing levels of capability.

10 Pickering, William H., with James H. Wilson. "Countdown to Space Exploration: A Memoir of the Jet Propulsion Laboratory, 1944—1958." In: Hall, ed., *History of Rocketry and Astronautics* (Springfield, Virginia: American Astronautical Society History Series, Vol. 7, Part II, October 1972).

11 WAC is thought by some to stand for "Without Attitude Control," others believe it stands for "Women's Army Corps," an allusion to the Corporal sister project. Pickering favored the latter.

12 Koppes, Clayton R. *JPL and the American Space Program.*

13 von Kármán, Theodore. *Pioneer in Aviation and Pathfinder in Space* (Boston and Toronto: Little, Brown and Company, 1967); Gorn, Michael H. The Universal Man: Theodore von Kármán's Life in Aeronautics (Washington and London: Smithsonian Institution Press, 1992).

14 Wilson, James H. Interviews with William H. Pickering. Pasadena, California: Oral History Program, Jet Propulsion Laboratory, 1972.

15 Pickering, William H., with James H. Wilson. "Countdown to Space Exploration: A Memoir of the Jet Propulsion Laboratory, 1944—1958." In: Hall, ed., *History of Rocketry and Astronautics* (Springfield, Virginia: American Astronautical Society History Series, Vol. 7, Part II, October 1972).

16 Beth Pickering Mezitt. Personal recollections of my Father, William H. Pickering. Private communication to the author: May, 2003.

17 Mudgway, Douglas, J. Oral history interview with William H. Pickering: Part 5. Pasadena, California, May 2003.

18 Koppes, Clayton R. *JPL and the American Space Program.*

19 Ibid.

20 Mudgway, Douglas J. Oral history interview with William H. Pickering: Part 5. Pasadena, California, May 2003.

21 Pickering, William H., with James H. Wilson. "Countdown to Space Exploration: A Memoir of the Jet Propulsion Laboratory, 1944-1958." In: Hall, ed., History of Rocketry and Astronautics (Springfield, Virginia: American Astronautical Society History Series, Vol. 7, Part II, October 1972).

22 Pickering, W. H. "Control and Telemetering for Corporal-E." Pasadena, California: Jet Propulsion Laboratory, Progress Report No. 4-15, May 1945.

23 Ibid.

24 Pickering, William H., with James H. Wilson. "Countdown to Space Exploration: A Memoir of the Jet Propulsion Laboratory, 1944—1958." In: Hall, ed., *History of Rocketry and Astronautics* (Springfield, Virginia: American Astronautical Society History Series, Vol. 7, Part II, October 1972).

25 Ibid.

26 Pickering, William H. and Robert J. Parks. "Guidance and Control System." Alexandria, Virginia: United States Patent Office, Number 3,179,353, 20 April 1965. Some time later, according to Pickering, the Army asked JPL to establish patent rights for this idea which, at the time, was being also being used by a contractor for the Army's Nike antiaircraft defense system. By establishing its prior use claim, JPL secured the patent thereby simplifying the Army's negotiations with its Nike contractor since JPL was already a prime contractor for the Army. "Unfortunately," he added, "we never collected any royalties on it."

27 Pickering, W. H., and J. A. Young. "External Instrumentation for NAMT." Pasadena, California: Jet Propulsion Laboratory Progress Report No. 10-1 to 10-9, 1946—1947; Pickering, W. H. and P. H. Reedy. "External Instrumentation for NAMTC." Pasadena, California: Jet Propulsion Laboratory, Progress Report No.18-1 to 18-8, 1949—1950.

28 Wilson, James H. Interviews with William H. Pickering. Pasadena, California: Oral History Program, Jet Propulsion Laboratory, 1972.

29 Mudgway, Douglas J. Oral history interview with William H. Pickering: Part 5. Pasadena, California, May 2003.

30 Ibid.

31 Ibid.

32 Ibid.

33 Wilson, James H. Interviews with William H. Pickering. Pasadena, California: Oral History Program, Jet Propulsion Laboratory, 1972.

34 Koppes, Clayton R. *JPL and the American Space Program.*

35 The drag brakes that Pickering refers to here were the outcome of the automobile "rear doors drag" experiment that the JPL team performed while driving back to the White Sands base from Las Cruces.

36 Pickering, William H., with James H. Wilson. "Countdown to Space Exploration: A Memoir of the Jet Propulsion Laboratory, 1944—1958." In: Hall, ed., *History of Rocketry and Astronautics* (Springfield, Virginia: American Astronautical Society History Series, Vol. 7, Part II, October 1972).

37 In those days, late-1940s to early 1950s, "upper-atmosphere" was synonymous with "space."

38 Pickering, W. H. "Study of the Upper Atmosphere by Means of Rockets." Pasadena, California: Jet Propulsion Laboratory, Publication No. 15, June 1947.

39 By international agreement, the IGY would run from 1 July 1957 through 31 December 1958.

40 The Redstone, a descendent of the V-2 German missile, was developed at the Army Ballistic Missiles Agency (ABMA) in Huntsville, Alabama.

41 Pickering, William H., with James H. Wilson. "Countdown to Space Exploration: A Memoir of the Jet Propulsion Laboratory, 1944—1958." In: Hall, ed., *History of Rocketry and Astronautics* (Springfield, Virginia: American Astronautical Society History Series, Vol. 7, Part II, October 1972).

42 See Folder 4 in the William H. Pickering Speech Collection. Pasadena, California: Archives and Records Center, Jet Propulsion Laboratory, JPL-181, 2003.

43 NRL in Washington, DC, was, in many ways, a counterpart to JPL in that it engaged in research and development tasks for the Navy rather than the Army.

44 See Folder 5 in the William H. Pickering Speech Collection. Pasadena, California: Archives and Records Center, Jet Propulsion Laboratory, JPL-181, 2003.

45 See Folder 6 in the William H. Pickering Speech Collection. Pasadena, California: Archives and Records Center, Jet Propulsion Laboratory, JPL-181, 2003.

46 See Folder 7 in the William H. Pickering Speech Collection. Pasadena, California: Archives and Records Center, Jet Propulsion Laboratory, JPL-181, 2003.

47 Mudgway, Douglas J. Oral history interview with William H. Pickering: Part 4. Pasadena, California, May 2003.

48 Beth Pickering Mezitt. Personal recollections of my Father, William H. Pickering. Private communication to the author: May, 2003.

Chapter 4

The Space Age Begins

Sputnik: Soviet Challenge

In the last weekend of September 1957, Bill Pickering made yet another trip to Washington, DC. Fall was a pleasant time to visit the capital. The oppressive heat of summer had given way to the milder temperatures of fall, and the frantic rush of tourists and vacationing school children had diminished to a quieter and more tolerable level of noise. The autumn tones of trees and gardens and the lengthening shadows of the waning sun gave the city a gentle patina of tranquility that would be a precursor to the harsh reality of the winter months to follow.

Pickering was a delegate to an international conference hosted by the National Academy of Sciences as part of its commitment to the U.S. participation in science programs for the International Geophysical Year (IGY). The IGY had begun on 1 July 1957 and was scheduled to run through December 1958. This conference, sponsored by the Comité Spéciale de l'Année Géophysique Internationale (CSAGI), would focus on planning the rockets and satellite part of the IGY program.

It was not the first time that the CSAGI had addressed this issue. Three years earlier at a 1954 conference in Rome, CSAGI had challenged IGY participants to give consideration " . . . to launching small satellite vehicles, to their scientific instrumentation and to new problems associated with satellite experiments, such as power supply, telemetering, and orientation of the vehicle." At that time, the Soviet Union had not joined the IGY program and only the U.S. responded.

By the time the CSAGI assembled in Barcelona two years later (1956), the Soviet Union had joined the IGY program and there its representative announced that the USSR would also use Earth-orbiting satellites to make measurements of temperature, pressure, cosmic ray intensity, micro-meteorites, and solar radiation during the IGY. He gave no further details or advance information. For its part, the U.S. undertook to fully describe the details of its sounding and satellite plans to assist those who wished to make correlated measurements by other means.[1]

Thus, by the time of the Washington conference, sufficient time had elapsed for both countries to implement their projects and throughout the week the meetings were pervaded by a general sense of expectation that an imminent launch date for an Earth satellite would be announced by either the U.S. or the Soviet Union. But the delegates were disappointed—there was no such announcement.

What happened next is best described in Pickering's own words:

> I was at that IGY meeting and on the Monday they had various nations get up and say what they were doing. The Russian delegate, Anatoly Blagonravov made the Soviets' report. He spoke in Russian and the translator followed him in English. Among other things, he said that they were getting pretty close to launching a satellite. The guy sitting next to me understood Russian and he said to me, 'He didn't say that—in Russian, he said the launching was *imminent*'.
>
> Now this was on Monday, and nothing happened until Friday when the meetings wound up and the Russians held a cocktail party to celebrate the end of the proceedings. So we all trooped over to the Russian Embassy for the cocktail party and there was the usual drinking and that sort of thing. Nothing was said about a satellite. Then during the evening, Walter Sullivan from the New York Times came in and asked me, 'What have they said about the satellite? Radio Moscow says that they have got a satellite in orbit.' So I talked to Lloyd Berkner who was the senior scientist present, and he and Walter Sullivan got Blagonravov aside and then Walter proposed a toast to the success of the Soviet satellite. Then the Russian vodka flowed like mad.[2]

Next morning a large gathering of scientists and reporters packed the auditorium of the National Academy of Sciences to hear Blagonravov give a lengthy public statement on the important new Russian achievement. While praising Sputnik as an example of Russian accomplishments in science and technology, he criticized the U.S. claims for its, as yet undemonstrated, satellite program. Nevertheless, the audience understood Blagonravov's pride in his country's achievement and applauded his assertion that this would "serve as an inspiration to scientists throughout the world to accelerate their efforts to explore and solve the mysteries and phenomena of nature remaining to be explored."[3]

The Russian criticism stung Pickering deeply. He knew, as few others did, that given the "go-ahead" a U.S. satellite could have been launched months earlier. Pickering recalled his frustration:

The reaction in this country was amazing. People were startled to realize that this darn thing was going overhead about ten times per day and there was not a thing they could do about it—and realizing that what was thought to be a nation of peasants could do something like this—with this amount of technical complexity.[4]

Speaking of the public reaction in this country Homer Newell wrote:

How brightly the red star shone before all the world in October 1957! Streaking across the skies, steadily beeping its mysterious radio message to those on the ground, Sputnik was a source of amazement and wonder to people around the globe, most of whom had no inkling of what was about to happen. In the U.S. many were taken aback by the intensity of the reaction. Hysteria was the term used by some writers, although that was doubtless too strong a word. Concern and apprehension were better descriptions. Especially in the matter of possible military applications there was concern, and many judged it unthinkable that the U.S. should allow any other power to get into a position to deny America the benefits and protection that a space capability might afford. A strong and quick response was deemed essential.[5]

Pickering returned to Pasadena to confront a sense of subdued frustration at JPL. The staff was, for the most part, well aware that the addition of a single live-stage motor to their existing upper-stage rocket motors could put a satellite in orbit. Furthermore, they knew how to do it and they had all the hardware they needed to do it. All they lacked was the approval to "go ahead." "We thought," said Pickering, "if the Army would only tell us to go ahead we could do this. We've got this reentry test vehicle—all we have to do is to put another stage on it and it will go into orbit." But the word to "go ahead" did not come.

Pickering was swamped by calls from the media wanting his reaction, opinion, and future predictions about the Sputnik affair but constrained by his orders from General John B. Medaris, he could only remain silent about his innermost thoughts for the future.

Despite its frustration over the rejection of the Project Orbiter proposal, and quietly encouraged by Pickering, JPL engineers began to formulate a quick response to the challenge posed by Sputnik. It would have to be something that the Army would support and for which it could get approval to carry out, and it would have to be something they could build quickly using material and parts that they already had. There would be no time for lengthy development and testing. The question was: what sort of response?

Pickering reasoned that, since the Soviets had more lifting capacity than the U.S., putting up a little satellite would not make much of an impression compared to Sputnik. Rather than just putting up another Earth-orbiting satellite, Pickering suggested they aim to shoot something to the Moon. "That would be dramatic, and a step in the right direction," he said. Calculations soon showed that if they put some larger upper stages on the Army's Jupiter,[6] employing much the same philosophy as they had used for Project Orbiter but on a larger scale, then they could indeed put a significant payload on the Moon.

That was it—they gave it a name, "Red Socks," and Pickering took it over to DuBridge at Caltech for his approval. "This is something where we can react to what the Soviets have done by doing something more dramatic," he said. DuBridge agreed and together they went to Washington to sell the idea to the Army. Although General Gavin, Head of Research and Development (R&D) for the Army, liked the idea, the Under-Secretary for R&D in the Department of Defense (DOD) deferred his opinion until an alternative proposal could be developed by the Air Force. So, in an atmosphere of inter-service rivalry, the Red Socks proposal foundered.[7] As far as the Pentagon was concerned the country had made a choice for its Earth-satellite program, and that was to be Project Vanguard.

Explorer: America's Response

About one month later a second Russian satellite, bigger than Sputnik 1 and carrying a live dog, appeared over the skies of the U.S. This second, spectacular Russian coup finally prompted the DOD to give the Army and JPL its long deferred authorization to prepare Project Orbiter as a back-up for the Navy's ailing Project Vanguard. It was the moment that the Army Ballistic Missile Agency (ABMA) and JPL had been waiting for since 1954.

Medaris called Pickering to an urgent meeting at Huntsville to inform him of the plan for implementing Project Orbiter. Pickering took two of his senior managers with him to meet with Medaris and the von Braun team. Pickering recalled the subsequent events:

> Medaris' office was right next to the Conference Room and I went in before the meeting and . . . basically I told him, 'You give us the Redstone and we will do all the rest.' He listened but didn't say anything—but when we went into the meeting Medaris said, 'Now this is the way it is going to be. . . .' And that was the way it was.[8]

ABMA would build the Redstone launching rocket to do the heavy lifting off the launch pad, and JPL would build the two upper stages and the satellite itself plus the instrumentation to go in it. JPL would also provide the radio telemetry and ground tracking system plus whatever guidance was needed to orient the spinning upper stage once it separated from the Redstone.

For some time, Pickering had been concerned about the future direction of the Laboratory mainstream effort. Although it was currently engaged with the Sergeant and final stages of Corporal as well as the reentry test vehicle (RTV) programs for the Army, he did not see a future role in military programs for the Laboratory. He was convinced that the path to the future began with a satellite program, and he was determined to lead the Laboratory in that direction. Here was the opportunity he needed to take the first step, and he intended to take it without hesitation. It turned out to be a seminal decision.

Work on the new project began immediately. Pickering assigned Jack Froelich to lead the project. Under Froelich's urgent direction, engineers began to assemble the scaled-down Sergeant motors that would provide the upper stages for the Redstone and to fabricate the satellite and the 4th stage motor that would finally inject it into Earth orbit. In a parallel stream of activity, Pickering's telecommunication expert, Eberhardt Rechtin, began to set up the radio receiving stations that would track the satellite and record its telemetry down-link, as it passed overhead. Suddenly the Lab was seething with energy and the spirit of incentive that had begun to dwindle under the dull pressure of the Sergeant program began to rekindle.

Pickering himself undertook the task of finding a suitable scientific package for the satellite. As a member of the IGY satellite-planning group, he was well aware of the science payloads that had been selected for the Vanguard satellite flights. One of them, Pickering suggested, Van Allen's Geiger-counter package, would be a logical choice of instrument for the cylindrical spinning body of Project Orbiter. The IGY committee concurred and directed Pickering to contact Van Allen and work out the details. Delighted with an opportunity to extend his cosmic ray research into regions beyond Earth's atmosphere, Van Allen agreed and arranged to provide the cosmic ray instrument and a scientist to help integrate it into the new Orbiter satellite.

A few weeks later, time ran out for Project Vanguard. The launch had been set for 6 December 1957 at Cape Canaveral, Florida, with full media coverage including national television. Public interest was very high, not only because of the partisan feelings aroused by the Sputnik affair two months earlier, but also because this was to be the first public viewing of a live rocket launch. Now the time had arrived. The countdown proceeded smoothly toward its spectacular climax. At T-zero, the gleaming slender launch rocket roared to life, spewed forth a giant plume of flame and smoke, then rose a few feet from the pad and exploded in a huge conflagration of orange and red flames. It was an apocalyptic moment of total failure before the incredulous gaze of a national audience. Unsympathetic press reports dubbed it the United States' "Flopnik."

For those new to the field of rocket development it was a moment of utter despair. For others, like Pickering and von Braun whose long experience in the field had made them aware of the difficulties that were associated with such complex systems, it was not altogether unexpected. They knew only too well that it took a long period of developmental experience, often painful, to achieve the exacting levels of reliability that such ventures required. That took time and experience, the one thing that Vanguard had not enjoyed.

For William Pickering and Wernher von Braun, however, there was another message in that dismal event—they were up next. The launch, of what was to be known for security purposes as Redstone Missile Number 29, was scheduled for 29 January 1958, a date that had allowed just 80 days from the order to proceed to launch. If the launch failed it could be attributed to a test failure of Army Missile 29; if it succeeded it would be hailed as America's answer to the Soviet challenge for space supremacy. Until then the payload for Missile 29 remained unnamed. The stakes were high for all involved.

Froelich's engineers had made good progress in preparing and testing their hardware for the January launch. In its final form, the satellite consisted of a steel cylinder, 80 inches long and 6 inches in diameter that contained the final 4th-stage rocket motor and the scientific instrumentation package consisting, principally, of Van Allen's Geiger counters. Two battery-powered transmitters would radio the science measurements to the ground and at the same time provide a signal for the ground receivers to track the position of the satellite as it moved across their field of view.[9]

Ever since he had demonstrated the efficacy of radio-based position finding techniques as compared to optical-based methods in the early Corporal guidance tests at White Sands, Pickering had maintained a strong interest in ground-based radio tracking techniques based on the Doppler principle. He believed that a velocity vector, represented by the Doppler component of a radio signal received from a moving missile or satellite, offered a more fundamental, accurate and practical data type than the angle data generated by optical devices for the determination of trajectories and orbits respectively. Pickering had charged Eberhardt Rechtin, another of his brilliant former Caltech students, with responsibility for developing a sensitive ground-based radio tracking system that could not only detect very weak signals from a tiny in-flight transmitter but could also extract the telemetry data that they carried. In addition, and perhaps most astonishing of all, it would use a small but significant part of the residual signal to measure the Doppler effect, or change in radio frequency, as the missile or satellite moved along its path relative to the receiving station. They called the system "Microlock," after the basic "microwave phase-lock" principle that enabled the sensitive ground receivers to track the phase of the satellite's incoming radio signals with extraordinary precision.

Microlock receivers at Cape Canaveral had been used for the early reentry test flights but now, faced with the need to cover an Earth-orbital flight, Pickering ordered his communications experts to set up three additional stations spaced roughly equally around the globe: one in California (near San Diego), one in Malaysia, and a third in West Africa. With the suitcase-size receiving equipment and a couple of technicians to set them up and operate them, Pickering had, in effect, a primitive international network of tracking stations for his satellite. These simpler stations would capture telemetry from the passing satellite, while the more complex stations in Florida and California would generate the Doppler data needed to determine the satellite orbit, in addition to capturing the telemetry data.

As the launch date drew closer, Medaris instructed Pickering to treat Missile 29 and its payload with the utmost security to keep its real identity hidden from public knowledge until after the launch. Decoys were to be arranged to cover the identity of all personnel that were associated with the satellite work, particularly Jack Froelich and Pickering himself.

As part of this subterfuge, Pickering planned to be in New York on 29 January, the opening day of the launch window, to present a paper at a meeting of the Institute of Aerospace Science. After the meeting, he planned to return unobtrusively to Washington to await the outcome of events at Cape Canaveral.

William Pickering did indeed spend that evening in Washington far from the unfolding drama at Cape Canaveral; not in the comfort of a quiet hotel but huddled over a single telephone in a small, nondescript conference room deep in the Pentagon, in Washington, DC. Earlier, the Secretary of the Army had "invited" Pickering, von Braun, Van Allen, and Berkner to follow the launch activities from there, probably to ensure that they were well hidden from public view. The office-size conference room contained a table and a few chairs, one standard telephone and, in the corner, a teletype machine that from time to time chattered with messages from the launch site control center, nothing more. To find out what was happening they would call the Cape, or vice versa.

Enormous sighs of relief and applause followed each successive announcement that the launch was "good," that the upper stages had separated, and that the high-speed cluster had fired. Finally, von Braun remarked, "the rest is up to JPL."[10] Pickering took over the phone to talk to Al Hibbs, his orbit determination expert at the Cape who was busy figuring out with pencil and paper just when the new satellite should appear over the horizon in California. It was then that the group was informed that, by an edict of the White House, the name of the new satellite was to be Explorer 1, and that there would be no public announcement until its signal had been picked up, and confirmed, in California.

Time seemed to stand still while they waited. The contact time predicted by Hibbs passed with no report of contact. Pickering recalled:

> So I am sitting there with the telephone stuck in my ear, and there is no signal, and the Secretary and all the others are glaring at me thinking, 'Where the hell is it?' I am really in the hot seat. So I tried to make chit-chat to Frank Goddard at JPL when suddenly he got a 'signal received' report from the nearby San Gabriel Valley Amateur Radio Club followed almost immediately

by a report from the San Diego Microlock. It was eight minutes late[11] according to Hibbs' estimate which, considering the real time nature of all this, was a pretty damn good estimate.[12]

A few moments later, a man from the Naval Research Laboratory (NRL) came in to confirm that a Navy station in San Diego had also received and confirmed the signal. There was much back-slapping, hand-shaking, and congratulating all around, but no champagne in the Pentagon office that night. Explorer 1 was in orbit, Pickering and von Braun were jubilant, and the nation was about to "go wild."

A short time later, Pickering, Van Allen, and von Braun were bundled into a car and driven a few blocks to the Academy of Sciences building over on the National Mall. The weather had turned rainy and cold and the streets were deserted. Here and there an occasional taxi, waiting impatiently at a traffic light, was the only sign of life in the sleeping city. Apparently the news that Explorer 1 was in orbit had preceded their arrival. A packed news conference, hosted by Richard Porter, Chairman of the IGY panel on Earth satellites, was in already in progress and news reporters and other interested observers of history in the making waited expectantly for further details. The crowd greeted their arrival, via a back entrance because they could not get in the front entrance for the crush of people, with great enthusiasm, and bombarded the three heroes with questions until well into the early hours of Saturday morning. An Explorer model happened to be on hand and pictures of Pickering, von Braun, and Van Allen holding the new satellite aloft were taken and subsequently published around the world.

It was no coincidence that an Explorer was available in Washington that night. Pickering observed:

> The plan always visualized success and that we were going to have a press conference and tie it into the National Academy of Science to emphasize that this was a scientific program, not just a stunt or a military program. Also, the committee that approved Van Allen's payload was essentially an IGY committee so in that sense it was tied to the IGY. But as far as the general public was concerned, the whole thing was perceived as just a reaction to Sputnik.[13]

The next morning the news press, many with extra editions, announced the achievement with blazing headlines: "U.S. SATELLITE RINGS EARTH: Army Launches Moon into Space," (*Los Angeles Times*); "U.S. SATELLITE CIRCLING EARTH: Caltech Moon Launched," (*Los Angeles Examiner*); "ARMY LAUNCHES U.S. SATELLITE INTO ORBIT: President Promises World Will Get Data," (*New York Times*). Pictures of Pickering, von Braun, Van Allen, and members of the JPL team and details of the Redstone and Explorer filled the back pages. Radio and television broadcasts scrambled to get the news on their early morning news programs.

Pickering, Van Allen, and von Braun show Explorer 1 to the world (NASA Image P8485).

Articles on JPL, and profiles of Pickering, von Braun, and Van Allen appeared in many of the prestigious newspapers and magazines around the world. Overnight, it seemed, Bill Pickering had been thrust into the harsh spotlight of public attention and his image, together with that of von Braun's, came to represent an icon for America's venture into an awesome new frontier beyond the familiar boundaries of Earth—the frontier of space.

When Pickering returned to Pasadena the next day, he stepped into a different world from the one he had left a few days earlier. Newsmen seeking interviews and comments, calls of congratulation and organizations requesting appearances poured into his home and JPL office from all sides. He had, in a word, become a "personality." As he recollected, "it took a bit of getting used to, because so many people were wanting to say a few words or take a picture. You were on the spot all the time."[14]

His children too, became objects of attention when their teachers explained the significance of the event and its association with their local school. At last, Muriel understood what her husband had been doing for the past several years. She began to share some of the glory and for her too, life was never the same. Now she was expected to appear with her husband at public functions to recognize his achievement and to fulfill her role as a part of his newfound public image.

The four remaining launches in the Explorer project played out over the next few months with mixed success. Two were highly successful, and together with Explorer 1, returned important new science data that led to the detection and exploration of the great belts of radiation around Earth appropriately named the "Van Allen Belts," for their discoverer, James Van Allen.[15] The second and fifth launches failed.

In reflecting on the fortuitous outcome of the first Explorer, Pickering viewed it as the beginning of "a slow climb up the learning curve," where failure was not acceptable under any circumstances, and the only path to success lay through costly experience, thorough understanding of the total problem, and exacting attention to the minutest of technical detail. This early insight of the problems inherent in the new technology of space flight, would serve him well in the stressful years that lay ahead.

A Very Public Figure

Pickering's speech to the Institute of Aerospace Science in New York in late January 1958 marked the last of his public appearances as a "private" figure. Forty-eight hours later that changed dramatically with the success of Explorer 1 and he became a very prominent "public" figure in every sense of the word, nationally and internationally. His public utterances on "space," and everything remotely related to it, appeared in the news media, radio, and television across the country and around the world; requests for personal appearances throughout the country poured into the JPL director's office. Undaunted by the extra workload and traveling it entailed, Pickering did his best to accede to most of the requests. He came to regard his public speeches as "part of my job," and used them to advocate his opinions on a variety of less technical subjects related generally to the U.S. space program and the Soviet threat to U.S. technological superiority. His public speeches, which were about to become a dominant part of his professional life, reflected his changing outlook, the breadth of his interests beyond the purely technical, the depth of his insight, and the public spirit of the times.

Borne on the tremendous wave of public interest in the new technology of "space," that followed the Explorer and Vanguard launches, Pickering addressed the Los Angeles Chamber of Commerce in March 1958. His topic was "The Engineer in the Space Age."[16]

He said:

> This meeting [today] is another evidence of the tremendous interest which the public displays toward this newest technical break-through-the dawn of the Space Age . . .The progress of civilization is today measured to a large extent by its technological achievements. . . .

Acknowledging that America's technological leadership had been challenged by the Soviet launching of Sputnik and recognizing that within five years the Soviets could challenge the U.S. in all fields of science, Pickering proposed a policy for the future that would reverse that imbalance and strengthen the United States' response to the Soviet challenge. He believed that "the answer lay with our scientists and engineers and with the support which we, as a nation must give them." He called for "a better understanding on the part of our political and industrial leaders of the importance of science to the national welfare." More support for basic research, more scientists and engineers, and more and better teachers to train them were the essentials of his message. He asserted:

> University and high school teaching must emphasize the real essentials of science and the scientific method. Mathematics and physics must become of prime importance. The ability to think clearly, to analyze a problem on the basis of essential data, to understand the fundamental principles involved these are the essential skills . . . and teaching is a critical factor, and Russia is doing a better job than we are.

The need for more scientists and engineers, higher standards for advanced education, better-trained teachers, and a national space program under civilian control: these were recurring themes in many of his speeches that followed.

While satellites, rather than manned space flight, were obviously of more immediate interest to Pickering, he had very clear ideas about the role that man should take in space. He gave voice to those ideas the following month in Denver, Colorado, at a symposium on "Man in Space" sponsored by the Air Force Office of Scientific Research.[17]

He began with the assumption that " . . . we have the capability to place a man in space in an Earth satellite or on some extraterrestrial mission," and then posed the question, "What do we gain by placing a man in the vehicle?" It depends, said Pickering, on the purpose of the mission. If the intent was to land a man on the Moon, or Mars, then a human passenger was obviously required. If however, the objective of the mission was to make scientific observations then, Pickering asserted, remotely-controlled instruments were a better alternative.

For scientific missions he saw the human passenger as " . . . an unnecessary complication."

However, when it came to the actual exploration of the Moon or planets, Pickering said the task of navigating across the vast distances of space and surveying the planet for viable landing sites would be done by robot vehicles, "but detailed exploration after a landing . . . must surely take man's intelligence."

In a prescient view of the distant future, Pickering summed up his thoughts on the role of man in space: "The capability for manned space flight becomes useful only when we consider the exploration of other planets. Before that time comes, unmanned vehicles can accomplish almost all of the missions assigned to space flight, in a cheaper more reliable fashion."

As Congress moved toward embracing a national space program, Pickering became concerned about potential conflict between military and scientific institutions for control of the nation's space enterprises. He voiced his concern in an address to the Association of the United States Army at San Pedro, California, in May 1958.[18]

Noting that the success of America's first Earth satellite was the result of a joint effort of the military and scientific communities, represented by ABMA and JPL, and that more joint programs were planned, Pickering said " . . . even as these preparations go ahead, unresolved questions cast their shadows before them . . . the essential ambivalence of the scientific-military mission in space will . . . become larger as the space program grows more ambitious."

He observed that while the scientist and the military man agreed on the necessity of exploring space and had worked closely together in the past to establish and demonstrate the basic principles, they were motivated by different objectives that could best be served by a separation of the programs in the near future. Pickering believed that these ideas were reflected in President Eisenhower's recent (2 April) recommendation to Congress " . . . to establish a new, independent federal agency that would be responsible for space technology, space science, and the civil exploration of space."

Congress was then considering this legislation; "I trust that a decision will be forthcoming before the end of the session," he said.

A week later, Pickering was speaking before the Louisiana State Department of Education in New Orleans. This time, the subject was "Education in the Space Age."[19] Pickering put his audience on alert when he opened with:

> There should be no doubt in the mind of any well-informed citizen that the U.S. and the USSR possess the power to annihilate each other, and this is a new situation which has never been faced before in the history of mankind. . . . Only seven months ago, on 4 October 1957, the U.S. lost an important battle in this conflict. The launching of the first Sputnik came as a shock to our people, but to people all over the world it was taken as proof that our much vaunted technological superiority was now second to the USSR . . . it was even interpreted to mean that the communist system had proved superior to the capitalist system. Fortunately, the launching of Explorer 1 on 31 January has done much to restore the balance. But it may be years before we have wiped out the memory of that October day.

"What do we, as citizens, do about it?" Pickering asked. "A few of us are in the front lines, so to speak, and we will fly satellites and lunar vehicles but we have no illusions about the strength of our opponent." Comparing the size and capabilities of the Soviet satellites with those of the U.S. led him to believe that the Russians were capable of "building, launching, and guiding rockets capable of being used as inter-continental ballistic missiles." "This is no time for complacency," he said, "We need to give all the support we can to our missile and satellite programs."

Pickering said that it was important to understand the true implications of the Cold War, the forces that controlled the political climate, the weapons being used, and the importance of science and technology. "We live in a time of tensions that are unlikely to be resolved in the next few years, and we must see that the coming generation is prepared for its part in the [continuing] struggle. The education system of our country must accordingly be prepared to accept the responsibility of training the citizenry for its part in this world struggle," he said.

Here, Pickering reiterated the themes that he had discussed at his earlier speech in Los Angeles on the "Engineer in the Space Age;" more engineers and scientists, better qualified teachers, higher standards for advanced education, well-rounded academic programs, and emphasis on mathematics and physics.

"For our very survival it is essential that the nation support a public education program adequate for the space age," he concluded.

As the euphoria over Explorer subsided and the sometimes wild conjecture about the future direction of a space program engaged the public interest, the reason for, and ultimate benefit of, a national space program came into question. With this in mind and, always alert for a catchy title for his speeches, Pickering chose "The Four Reasons" for his address to the Pasadena Chamber of Commerce in June 1958.[20] He said:

> We are standing tonight at what might be called the door to space . . . and we have paused momentarily on the threshold to collect ourselves and our equipment for the adventures that stretch before us. . . . There are two basic questions that can be asked at this stage—why do we want to go into space and how do we propose to get there?

Citing the President's Science Advisory Committee for an answer to the first question, Pickering said that each of the following four factors was reason enough to justify a national space program:

> Natural curiosity leads man to try to go where no one has gone before; being strong and bold in space enhances the prestige of the U.S.; A defense objective ensures that space is not used to endanger our security; and Space technology offers new opportunities for science experiments that will increase our knowledge and understanding of the Earth, solar system, and the universe.

"Taken together," he said, "they do indeed constitute a compelling reason to embark on such a program with all reasonable speed."

He explained how the development of large and powerful rockets for military purposes had in fact "opened the door to space," and in doing so had created a strong rivalry between our military and scientific institutions for control of the nation's space program. Although the present national space program was being directed by the Advanced Research Projects Agency (ARPA) of the DOD, Congress was considering a bill to establish a civilian agency, the National Aeronautics and Space Administration (NASA), with authority to direct "aeronautical and space research sponsored by the government."

How would these two agencies interact? Pickering believed

> The right answer must lie somewhere between the two extreme views. The costs of space vehicles are so large as to require a coordinated national effort. Therefore both military and scientific objectives must be considered . . . ARPA and NASA must cooperate so that all phases of our space program are put into proper perspective . . . we need a positive program, not hysterically reformulated every time Russia sends up another Sputnik, but logically and scientifically planned and funded with the objectives of reestablishing America's preeminence in this area of technology and insuring our military position in space, and opening a new era in human development.[21]

Although the U.S. program roughly paralleled that of the USSR, he believed that it lagged the Soviet program by about two or three years. He said:

> With the Soviets possessing this lead, it might seem that we will remain in a sorry state to contest their leadership. This, I refuse to believe. Admittedly we are a late starter in the race; most of our actions have been merely reactions to Soviet action that were motivated, to a large extent by fear, but surely this nation can cast off such a miserable, defensive psychological response. Here we stand at the door to one of the greatest adventures of the human spirit. We have an opportunity to show the world how we can lead mankind in hope and freedom, leaving behind fear and suspicion. We can do it—we must do it.

The Pasadenians rewarded his presentation with a long and enthusiastic standing ovation.

The first year of the Space Age made an arresting and appropriate subject for Pickering's address to the Aircraft Industries Association in November 1958. "Just a year earlier," he reminded his audience, "the country was still recovering from the shock of seeing the Russians place two satellites in orbit only about one month apart." Pickering said there had been a universal, almost hysterical demand that something be done quickly to off-set that perceived "Cold War victory." Now it was time to review what had been done to catch up with the Russians in the field of guided missiles and space development.

He carefully explained what was then known about Russian progress and what could be deduced from the performance of their satellites and ICBM test vehicles that had been launched successfully.

By comparing the size and weight of the Russian and American satellites launched so far, and the energy required to place them in orbit, Pickering observed that "the missile that launched Sputnik III must have been at least as large as America's most advanced heavy rocket, the Atlas. Yet the United States had not yet fired a complete Atlas, let alone developed it to the state where it could be used for a satellite launching vehicle." Obviously, Russia was "way ahead of the United States in the guided missile art."

In answer to the important question of "are we showing signs of catching-up?" Pickering could give only a qualified "Yes." He believed that a "psychology of failure" was developing in the nation, largely as a result of the public tendency to "gloss over" our failures in the missile and satellite test programs.

He cited recent media examples from the Vanguard, Explorer, and Pioneer missions that particularly annoyed him. "Better luck next time" or "Man against impossible odds," or "Man against Newton's laws." These were not examples of "glorious failures." None were acceptable. He asserted that these failures were due to "insufficient engineering analysis and test of a condition that was encountered in flight." He totally rejected a recent media comment that " . . . nothing succeeds like a failure." It was a carry over from the Vanguard experience and overlooked the point that a missile was a very complicated device that called for thorough engineering analysis and preflight testing. He said:

> We have tried to run before we can walk. . . . The key to obtaining real improvement [in the missile program] is a recognition by both the government and industry of the fact that missile design is an engineering problem, and that failure can mean one of two things, either a failure to understand the problem or poor engineering design. In either case, the lesson to be learned from failure is to review the problem in an analytical manner, then apply the results of analysis.

He blamed the publicity associated with the satellite programs for much of the problem. It was essential, he said, that our space programs not ignore the possibility of failures, but be conducted "in such a way that success is expected" and is not perceived "as a lucky fluke."

"We must set success as our goal and be content with nothing less. Only then will we catch, and surpass, the USSR in this vital area," he concluded.

As the space program moved forward, now under the direction of NASA, Pickering would have cause to reiterate the ideas expressed in this speech time and again, when a succession of temporary failures obscured the larger picture and threatened to divert attention from the intermediate steps that, in his opinion, were essential to the achievement of ultimate success.

The age of space had arrived.

Endnotes

1 Newell, Homer E., *Beyond the Atmosphere: Early Years of Space Science* (Washington, DC: NASA SP-4211, 1980).

2 Mudgway, Douglas J., Oral history interview with William H. Pickering. Pasadena, California, May, 2003.

3 Newell, Homer E., *Beyond the Atmosphere: Early Years of Space Science* (Washington, DC: NASA SP-4211, 1980).

4 Mudgway, Douglas J., Oral history interview with William H. Pickering. Pasadena, California, May, 2003.

5 Newell, Homer E., *Beyond the Atmosphere: Early Years of Space Science* (Washington, DC: NASA SP-4211, 1980).

6 At that time, the Jupiter was being developed as a 2000-mile range, intermediate range ballistic missile by the United States Army

7 Pickering believed that this decision was representative of the rivalry that existed between the Army and the Air Force.

8 Mudgway, Douglas J., Oral history interview with William H. Pickering. Pasadena, California, May, 2003.

9 Pickering, William H., with James H. Wilson, "Countdown to Space Exploration: A Memoir of the Jet Propulsion Laboratory, 1944-1958." In: Hall, ed. *History of Rocketry and Astronautics* (Springfield, Virginia: American Astronautical Society History Series, Vol. 7 Part II, October 1972).

10 Ibid.

11 Its late arrival over California was attributed to the extra velocity it picked up as it passed through the jet stream, which resulted in a slightly larger orbit than Hibbs had predicted.

12 Mudgway, Douglas J., Oral history interview with William H. Pickering. Pasadena, California, May, 2003.

13 Ibid.

14 Ibid.

15 Van Allen, James A., "Radiation Belts Around the Earth." *Scientific American*, 200, March 1959.

16 See Folder 12 in the William H. Pickering Speech Collection. Pasadena, California: Archives and Records Center, Jet Propulsion Laboratory, JPL-181, 2003.

17 Ibid. See Folder 14.

18 Ibid. See Folder 15.

19 Ibid. See Folder 16.

20 Ibid. See Folder 18.

21 Ibid. See Folder 22.

Chapter 5

The Learning Curve

Toward the Moon

As the Explorer program moved toward its conclusion in the fall of 1958, Pickering's dreams of sending a spacecraft to the Moon began to take form and substance in a project called Pioneer.

Directed by Advanced Research Projects Agency (ARPA), the program was shared between the Air Force and the Army with two probe launches assigned to each of the services. The probes would carry temperature sensors and the now-famous Van Allen Geiger counters to extend radiation belt studies deeper into space, hopefully to the region of the Moon. When the first two Air Force-managed Pioneers failed, the Jet Propulsion Laboratory/Army Ballistic Missile Agency (JPL/ABMA) team was quick to set early December 1958 for its Pioneer launch attempt.

Essentially, simplified versions of the earlier Red Socks proposal, the JPL/ABMA lunar probes were the third and fourth launches in the Pioneer lunar probe series. ABMA supplied a much more powerful launch vehicle called Jupiter to lift JPL's upper stages for injection on to a trajectory that would reach the vicinity of the Moon, and perhaps beyond. JPL supplied the conical satellite with its scientific instrumentation and a new, more refined radio tracking and telemetry system that operated at 960 MHz, a much higher frequency than that used for the Explorers.

Pickering knew that to maintain continuous radio contact as Earth rotated under a probe traveling on a trajectory in deep space would require a three-station network of receiving stations. This meant that the upgraded radio system now carried by Pioneer 3 would require a completely new, and much more complex, ground tracking system than the simple arrangement used for Explorer. To bring this into being on short notice he depended on his former student, and trusted communications chief, Eberhardt Rechtin. For the immediate purpose of tracking Pioneer, Rechtin needed a large, steerable,

26 meter-diameter, parabolic, antenna—a facility that, at the time, existed only in the minds of radio astronomers. Six months later, after a prodigious design, procurement, and construction effort, the huge antenna, the first of its kind, stood ready for operation at a remote site near Goldstone Dry Lake in the heart of California's Mojave Desert. As Pickering so well understood, this effort and what sprang from it, was quite as significant to the U.S. space program as the launch itself.[1]

When Pioneer 3 was launched three days later on 6 December 1958, the new Goldstone antenna had no problems in tracking the probe out to 63,500 miles, the limit of its flight, but regrettably far short of the desired lunar distance.

A few weeks later, as if to mock the U.S. effort, the Soviets announced the success of their Luna 1 space probe. It had passed within 4,000 miles of the Moon on 2 January, and continued into orbit around the Sun claiming the distinction of the first spacecraft to escape the clutches of Earth's gravitational field. The race was truly on, and no one felt it more keenly than Pickering.

Although Pioneer 3 did not succeed in reaching the vicinity of the Moon, the next attempt with Pioneer 4 in March 1959 was completely successful, and did much to assuage the disappointment of the earlier launch. However, it did little to assuage Pickering's aspiration to be "first in deep space." He sought a more ambitious space program that could clearly demonstrate the superiority of U.S. technology and management and secure a reputation for JPL that none could challenge.

In April of that year, Pickering gave an address to the alumni association of California Institute of Technology[2] that epitomized his innate desire to be first in deep space. Obsessed with the perceived relegation of U.S. to second

Pioneer 4 became the first U.S. space probe to explore deep space on 3 March 1959 when it passed the Moon at a distance of 37,300 miles and continued to gather space science data until its batteries became depleted at a distance of 407,000 miles from Earth. The fiberglass cone that enclosed the electronic package is 50 cm long and 25 cm in base diameter. The gold coating makes it electrically conductive and the white stripes serve to control its temperature in space to about 40 degrees centigrade. An 8 cm long probe at the apex is electrically insulated from the cone to create a dipole antenna for radio transmissions to Earth on a frequency of 960 MHz (Photo: NASA/JPL-Caltech Archives, Photo number 291-3730).

position in the space race, he burned with ambition to bring the U.S. program into the dominant position. In "The Exploration of Space" he made a plea for public support of the emerging U.S. space program: " . . . the public generally has not been made aware of the urgent significance of supporting a basic research program that will pay off some X years later with a visible launching of a rocket headed perhaps to the Moon or to the planets." He thought the public was justified in asking what had been achieved since the U.S. started with a rush to compete with Russia in space. In fact we have done a great deal, but most of it is below the surface, he said "we have added greatly to the knowledge about the Earth, its atmosphere and conditions in space, and paved the way with a Vanguard meteorological satellite for what someday we may be able to use as a weather satellite for long-range weather forecasts." On the other hand, to the scientists, the discovery of the Van Allen radiation belts was of major scientific interest. But we could not expect the scientific and non-scientific communities to be stimulated by the same things. The facts show " . . . that there are far more people interested in spectator sports in this country than are interested in the more esoteric aspects of science," he said, and " . . . this lack of interest in the motivation and meaning of science will become increasingly important as the budget demands for space experiments go higher and higher."

At present, public support for our space programs was being sustained by wounded national pride in the obvious success of the Russian rocket technology but, he asked, "What of the time when, and if, it occurs, when we achieve equality with the Russians. Will the American public lose its interest in space and demand reductions in the space budget?" Pickering thought that the future held much of great practical interest to the public. He cited weather satellites as an example, but believed that the most public interest would be aroused by the possibility of developing "communications satellites by which it would be possible to beam television programs from Europe to the U.S."

But first, he cautioned, to even think about catching up to the Russians, we must have bigger and better launch vehicles, or booster rockets, to lift the heavy payloads into space. The existing launch vehicles, based on military designs for ICBMs were simply not powerful enough. He noted a succession of heavy lift boosters that were currently under development for non-military purposes; the largest two were the Saturn and Nova. The Saturn would be capable of placing heavy payloads, including manned capsules, into Earth orbit, while the Nova would be " . . . capable of transporting a man to the surface of the Moon and returning him to a safe landing on Earth."

He said that these programs would be costly, and the public would be justified in asking for a reason why it is important to do these things and what it would expect to get out of it. "We do these things," he said, "because of the unquenchable curiosity of man. The scientist is continually asking questions and setting out to find the answers. In the course of getting these answers, he has provided practical benefits to man that have sometimes surprised even the scientist." He cited Roentgen's discovery of x-rays and Van Allen's discovery of the radiation belts as examples. "Who can tell what we will find when we get to the planets . . . or predict what potential benefits to man exist in this enterprise. It seems to me that we are obliged to do these things, as human beings," he concluded.

It was a bold statement, issued from the heart of a man who passionately believed in the issue he advocated, and who possessed the personal courage and technological acumen to back it up.

As evidenced by his public utterances in 1958 following the Explorer success, Pickering had given considerable thought to the form in which the U.S. should expand its interest in the new field of space, and he had formed very definite ideas about the role that he saw for JPL in such an enterprise. He believed that military interests and civilian interests in the development of space were sufficiently divergent to justify the establishment of entirely separate and independent programs to serve the needs of each. The civilian program would be the frontline challenge to the Soviet thrust into space, and should leapfrog ahead of the Soviets with bold attempts to reach Venus and Mars rather than diverting effort to reach the Moon. In Pickering's view, JPL's demonstrated expertise and experience clearly placed it in an indisputable position to lead such a program.

Meanwhile, down the corridors of power in Washington, DC, alternative plans to create a new government agency to handle the nation's space program were rapidly gathering form and substance.[3]

By the time President Eisenhower signed the bill that created the National Aeronautics and Space Administration (NASA) in July 1958, Pickering was resigned to the fact that he would have to find a place for JPL in the NASA organization if he was ever to realize his ideas for a national space program. Not all of his executive staff agreed with him. Some held little regard for the bureaucratic constituency of NASA and its aerospace-driven background. They feared that the talents of JPL would be squandered if it became merely a research "service" for the aerospace industry. Nevertheless, Pickering held to his own opinion, knowing that the JPL-Caltech hierarchy also believed that "the Laboratory had a mission to set the new space agency on the proper course."[4]

When NASA sought the views of the scientific community and specialists in relevant fields, as a basis for developing its space science program, Pickering lost no time in offering his point of view. Pickering's conceptual ideas for a space science program, and how they might be carried out, were reflected in a five-year plan of solar system exploration developed for the new Agency by a JPL team led by Albert Hibbs.[5]

In the firm belief that JPL should play a leading role in NASA, Pickering wrote to J. R. Killian, President Eisenhower's science advisor, setting out his view on the role that JPL should play in the new administration. He warned Killian that unless the new Agency accepted the concept of JPL as the national space laboratory, there would be a danger that the military would seize the initiative and leave NASA to provide only supporting research with an occasional scientific payload. However, as the national space laboratory, JPL with all of its unique experience would become a key NASA resource that, given a clearly defined responsibility, could draw up a viable long-term space program for the nation.[6]

Late in 1958, when NASA approached Caltech with the idea of transferring JPL from Army to NASA jurisdiction, both Pickering and Caltech President Lee DuBridge responded with enthusiasm. Under the final arrangement, Pickering agreed to complete the Sergeant weapon development program for the Army and to do some further limited research for the Army. On 3 December 1958, Eisenhower signed an executive order that authorized the transfer and Pickering, his entire staff, and JPL's future research programs came under the direction of NASA.

Thus, it came about that Pioneer 3, launched on 6 December 1958 and the new 26-m antenna at Goldstone that tracked it, combined to carry out NASA's first mission into deep space.

Early that morning, NASA convened its first post-launch press conference at its temporary headquarters in Washington, DC, with Dr. Abe Silverstein, Director of Space Flight Development representing NASA and Kellogg, von Braun, and Pickering representing the IGY Satellite Panel, ABMA, and JPL, respectively. Unlike the Explorer event, this press conference was very formalized and far less spontaneous and exuberant. It marked the first of many media events that exposed Pickering to public scrutiny. Some would be excruciatingly depressive, others wildly exuberant. Silverstein said the Pioneer 3 launching had been "functionally successful" and called the accomplishment " . . . a supreme achievement in the engineering sciences and the arts." He paid tribute to the cooperation of ABMA and JPL that had brought about this contribution to the IGY and noted that the Pioneer 3 program was managed by NASA under the direction of Dr. Keith Glennan. NASA was making sure that the public recognized the new order of authority in the nation's space program. Most of the questions from the audience were related to an explanation of how the rocket and upper stages worked; Pickering, von Braun, and Kellogg answered cautiously.[7] In conclusion, Silverstein extended NASA's thanks to "the Army Team, the team at our new Jet Propulsion Laboratory at Pasadena and to the scientists who provided the instrumentation that went into the Bill Pickering payload." Silverstein was bent on driving home the message that NASA was in control.[8]

Pickering, however, chaffed under the tightening constraint that he perceived NASA was imposing upon his personal ambitions for the exploration of space. The Explorers and Pioneers represented, like the first Corporals that went before them, an improvised solution to an existing problem that that made use of old ideas and, at best, current technology. Pickering envisioned much more ambitious and challenging goals for the embryonic space program of the U.S. But NASA regarded such ambitions with reservation and came to perceive Pickering and JPL as a significant, but irritating, member of its family of Field Centers.

At the end of September, as the Sputnik affair neared its second anniversary, Pickering spoke before the 1959 Annual Meeting of the American Institute of Chemical Engineers in St. Paul, Minnesota, on the subject of "Space—the new scientific frontier."[9] The nascent organization of NASA had begun to formulate its plans for a national space program and, although the role of JPL in the new Agency had not yet been fully established, Pickering had become reconciled to the fact that if it was to survive, JPL would have to accept authority and direction from Washington. Nevertheless, Pickering had very clear ideas about the direction the space program should take and plenty of advice for NASA as to how to go about it, as this speech clearly shows.

Pickering began by expressing his concern that the current wave of public support for the U.S. space program that had been energized by the Russian show of technological superiority would soon subside as the reality of the enormous costs associated with a viable space program became apparent.

Pickering used a comparison of successful space shots, eleven for U.S. versus three for USSR, to show that while the U.S. could point to more satellites and more, and better, science results, the Soviets had put much heavier payloads into space, including one that hit the Moon. Obviously this achievement implied the use of more powerful rockets with superior guidance systems. "Whether we like it or not," said Pickering, "this capability can, and presumably is being used to build missiles which place all of the U.S. cities within range of launch sites in Russia."

Pickering did not see any evidence that we were catching up with the Russians and believed that the nation " . . . must understand the nature of the task ahead of us and the way in which we must organize to accomplish the task." In a world where technological achievement is regarded as the mark of success of a civilization or a political system, satellites and Moon shots represent achievements of the highest order, as the Russians obviously realized. For much of mankind, exploration of the heavens represented an entirely new thought, and as much a landmark in the history of human development as Darwin's theory of evolution. Thus, Pickering reasoned " . . . space achievements become one of the most important weapons in the Cold War."

Pickering expressed the view that we needed a realistic evaluation of our future national space program that would allow us to

> . . . advance as rapidly as possible to a position where we have
> begun to explore the Moon and the nearby planets. . . . Then,

we can consolidate our scientific knowledge with detailed experiments, and then we can establish possible military applications. We should not now divert our efforts into costly military ventures of doubtful value. . . . Which was more important to the real interests of the United States, a military photographic satellite, or an interplanetary vehicle which could give an answer to the question of life on Mars?

Either one, he thought, could be accomplished in the next few years, but not both. "I happen to believe that the mission to Mars should be given priority," he said.

Pickering delivered a rousing finish to his address. The exploration of space, he asserted, presented us with much more than a fascinating new scientific frontier; it represented a new dimension in human thought and gave us a powerful weapon in the Cold War. He asked:

Where will we stand ten years from now? Will we still find the Russians scoring firsts in space? Will the Russian Prime Minister send our President a desk ornament made from lunar materials? Will the new map of the Moon carry Russian place names? If so, Khrushchev will have been right; Communism will 'bury Capitalism.'

"But this does not have to be true," he explained. "We have the resources to respond to this challenge. Give us your support. Try to understand our program. Separate the realities from science fiction–then stand back. Watch us go!"[10]

The JPL Director's plan for future exploration of space caused considerable comment when it arrived at NASA in April 1959. Known as the Hibbs report, the plan called for launching four spacecraft to Venus, three spacecraft to Mars, and five spacecraft to the Moon, over the period August 1960 through March 1964. The details of the report bore the unmistakable imprint of the ambitious Pickering ideas.

But NASA demurred. Pickering's vision of a planetary program focusing on the planets under his direction was finally dashed, when in mid-December 1959 he received a letter from Richard Horner, NASA's Associate Administrator, directing him to concentrate on lunar rather than planetary exploration. However, much of his private opinion differed from that of NASA; there was no option other than to move forward as directed. The free-wheeling Army days were long gone.

Sensing Pickering's displeasure at these instructions, Dr. Silverstein, Director of Space Flight Development at NASA, sent several of his top scientists to JPL to discuss the details of the NASA program with Pickering and his staff. In the less than cordial discussions that followed, Pickering, Hibbs, and several other top JPL scientists made their objections and concerns known to the NASA officials. When asked whether NASA had considered the question of competition with Russia, scientific objectives, and the matter of organization in its planning, Newell pointed out that the overall objectives of the NASA program in space flight were: the extension of the domain over which man may move and be active, and the extension of human knowledge about Earth, its environment and space, and the objects of space. NASA regarded both of these as very important and had designed programs to support them strongly. Referring to the matter of competition with Russia, Newell said:

> In the matter of Russian competition, it is clearly understood that whether it be stated openly or not, the United States is in competition with Russia and the stakes are very high indeed. It is further understood that the loss of the space race would be of great seriousness to the United States, economically, culturally and politically.

Pickering could have had no dissent from that point of view coming from NASA's top spokesman. Newell continued:

> But it is felt that our competition with Russia must be based on a sound program of science and technological development and not on the performance of what may be called 'stunt-type' missions. If the latter approach were taken, we would be in danger of being scooped or bettered by the Russians and made to look even worse than we are, and in the long run we would lose out by not properly developing our ability to compete.[11]

How those words must have resonated with Pickering as he heard his own opinions from a dozen speeches over the past year, echoed back to him from the voice of NASA.

Perhaps they softened his attitude toward his new masters for at the end of the negotiations, it was resolved that:

> . . . NASA Headquarters would remain responsible for overall program planning, while JPL would lead the engineering and execution of lunar and planetary missions—a position that it has maintained for the most part through the present time. NASA officials assured JPL that while lunar exploration remained the Agency's main area of solar system interest, planetary work would get underway soon, with launches to Mars and Venus whenever they were in optimum position for a planetary mission . . . and finally, NASA pledged to create a single working committee for lunar and planetary exploration in the NASA management structure."[12]

With the scope of JPL's responsibilities clarified, the parties hammered out a compromise that blended NASA's immediate interest in lunar flights with Pickering's longer term interest in planetary missions.

The first two lunar flights would be essentially engineering test flights to evaluate requirements for attitude control and communications. The final three flights would gather scientific data about the lunar surface in the "period immediately preceding impact." In one of the few documented cases where JPL and NASA reached a simple mutual agreement the overall project was named Ranger. The Ranger missions would demonstrate the ability to perform a scientific program in space and clarify the requirements for attitude stabilization and planetary communications—all of it technology necessary for its subsequent planetary missions to Venus and Mars. These later missions were to be named Mariners.

In his history of project Ranger, Cargill Hall wrote, "The Ranger program would also meet another need, publicly expressed by JPL Director William Pickering, to demonstrate the superiority of the 'American Way' to uncommitted states in the international community."[13] It was a very pertinent comment, as Pickering's public statements clearly showed throughout 1959.

In addition to his involvement in the formative discussions with NASA, then in progress between Washington and Pasadena, Pickering found time to deliver several public addresses in the last quarter of 1959.

His address to the Seventh International Meeting of Communications (Engineers) in Genoa, Italy, was a prime example of Pickering at his best, a lecture-style delivery with equations, anecdotes, illustrations, and technical wisdom and foresight and, for its time, a most compelling subject: "Communications with a Lunar Rocket."[14] He began by comparing the new challenge in radio communications technology—communicating over distances that had increased by four orders of magnitude—with that of the change in technology "brought about by the atomic bomb which increased the power of explosives by some such amount." Rockets will soon be available that can send spacecraft into orbits that extend far beyond Earth's orbit around the Sun, he said. It would be up to the communications engineers to make these vehicles useful by returning their data to Earth.

Pickering explained how the basic "radar equation" related the operating frequency, antenna size, transmitter power, and receiving capability of a radio system to the distance over which it could send and receive data, and illustrated his argument using examples from the recent Pioneer space mission and the new Goldstone antenna. He covered all the fine technical nuances that affect the performance of a communications channel to predict that "communications engineers will indeed be able to provide communications for space vehicles traveling far throughout the solar system." Finally, he pointed out that space vehicles that might take 8 to 10 years to reach their destinations afforded communications engineers an opportunity to improve their Earth-based techniques as the mission progressed, so that when the vehicle eventually arrived at its target, they would have the necessary capability to communicate with it.

"Do We Have a Space Program?" delivered to the American Rocket Society (ARS) in Washington in November was essentially a replay of the themes that he had espoused to the Institute of Chemical Engineers in September: the importance of a successful space program as a powerful weapon in the Cold War, U.S. versus USSR, seen as a clash in cultures, openness versus secrecy, do we have the right space program? [15]

Pickering repeated his strong belief that "our national stature and prestige in the world" was at stake in our race with the Russians. He reiterated his opinion that "in the 2 years since Sputnik, we have not succeeded in matching the Russian achievements."

The remedies that Pickering proposed were similar to those we saw earlier: make the public understand the importance of a space program, clearly define a national objective, establish management and funds to properly support it, and clarify the relative priorities of our civilian and military programs. And then he called for action: " . . . as professional engineers and scientists . . . our task is to educate the public and Congress to the realities and needs of a national space program," he said.

The prestigious space journal *Astronautics* reprinted the full text of Pickering's speech in its January 1960 issue.[16]

Pickering closed out 1959 with an address on "The Scientific Uses of Artificial Satellites and Space Craft" to the Association for the Advancement of Science in Chicago on 26 December.[17] Somewhat like his address in Italy on space communications, this speech was completely technical and addressed the scientific implications of the science program. He pointed out that there were four types of experiments that could be made with Earth satellites: radiation measurements, magnetic field measurements, observation of the appearance of the earth from space, and observations of residual atmosphere. In addition, information about the gravitational field and the shape of Earth could be obtained from precise observations of satellite orbits. Many of the experiments had already been carried out and had given startling results, most notably the discovery of the Van Allen belts of radiation that surround Earth.

Pickering saw a bright future for science in space. He concluded:

> Space vehicles that come close to the Moon or planets and eventually land softly on the Moon or planets open up a whole new era of scientific exploration that covers all of the natural sciences. . . . The discovery of life on some planet will be a most important factor in answering the question of the origin of life, and be as important a factor in human thought as the theory of evolution.

Pickering appended an interesting table to this speech. It compared U.S. launchings and successes with those of the Soviets. The numbers spoke for themselves. It had taken the U.S. 27 attempts to get 13 satellites into Earth orbit successfully. In the same period, the Soviets had succeeded six times with an unknown number of attempts. Pickering believed it unlikely that they had achieved a better success rate than the U.S. He asserted that these numbers were an encouraging indication that the U.S. was rapidly catching up, despite the public's perception to the contrary.

William Pickering turned 49 years old in December 1959 just as these momentous ideas were becoming a reality. Still a young man by any measure, ambitious, gifted, and assured of his place in the technological world of his choice, he brought his talents and a sense of hubris to NASA as later events would clearly show. But for now, he had come to terms with the inevitability of the new direction imposed by NASA upon the Laboratory, and whatever private reservations he may have had about the scientific wisdom of NASA's lunar program he held no illusions about its importance to the international prestige of the U.S. NASA's promise of later missions to Venus and Mars provided additional incentive for his commitment to the Ranger and Mariner programs. Between them, the Ranger and Mariner programs would dominate the rest of his professional life at JPL and determine the fortune of the thousands of young men and women, scientists, engineers, and support personnel upon whose unique skills and dedication the success of both programs depended.

The Learning Curve

At the beginning of 1960, the course for JPL's foreseeable future in space had been essentially determined. Under NASA's direction, JPL would carry NASA's lunar program forward with Project Ranger, while the parallel Mariner program would represent NASA's interest in planetary exploration, beginning with Venus and Mars. Pickering regarded Mariner as the more important program in terms of enhancing the "national prestige" ethos to which he so strongly adhered. He viewed the Ranger program, essentially, as a means to gain access to the "learning curve" of experience in designing and building planetary spacecraft and operating them in deep space.

Pickering organized the technical staff of the Laboratory in the form of a matrix: technical divisions vertically and flight projects horizontally. Each flight project office drew on the technical talent available in each of the divisions, as required to support its individual project. At the completion of each project, the assigned engineers resumed duty with their line divisions. In this way the best talents in each discipline were available to the flight projects for as long as needed to complete each task.

For Pickering, the "matrix" was yet another source of contention with NASA. NASA believed that a hierarchical organization, rather than a matrix arrangement, was much easier and cleaner to manage for carrying out project-type enterprises. "But I wanted to hang on to the matrix form," he said, "because the projects have a relatively short life and there are a multiplicity of them. So you should be able to pick people out of the matrix and have them work for two bosses, one in the Project, the other in the Division. Furthermore, you can use other [experts] in the matrix to help solve problems [that arise] in the projects."[19] Obviously, to make that idea work, the project manager must have commensurate authority. Pickering recognized that fact and chose very strong personalities for his project managers and gave them a direct line of authority to himself.

Pickering depended largely upon weekly meetings with his senior staff, a group of about 30 of the Laboratory's executives and top level managers from the technical and administrative divisions, to keep track of progress on the flight projects, discuss problems and their resolution, and to discuss and disseminate policies and practices within the Laboratory.[20] The weekly Director's meeting became the forum for interaction between Pickering and his top-echelon executives:

> I did not try to enter into the day-to-day decision making for example. That was one of the strong points about the organization, we allowed people throughout the organization to talk to one another and do things as they saw fit without having to come all the way to the top. I was not involved in day-to-day thinking. When they had problems, the project manager dealt with it. My interaction was to isolate the in-house Ranger project from NASA, and eventually from Congress, of course.

He explained ruefully that, although NASA was the primary contact with Congress, the fact of the matter was that "Congressional committees wanted to come out here and visit the place and talk to various people like myself, so that I ended up getting more and more involved in appearing before Congressional committees to talk about these things."[20] Congressional and other high-level visitors were soon to demand a significant portion of Pickering's on-lab time. Always the most courteous of hosts, Pickering found few of the visits of significant value to his program except in the most general way. But he said that was part of his job.[21]

While Pickering accepted the inevitability of NASA's direction of JPL, that fact in no way reflected his perception of the big picture. The government's delineation of the rationale for a national space program and its designation of responsibility for carrying it out, remained issues of great contention to Pickering in early 1960, and he took advantage of every opportunity to make his views known to the public as evidenced by this speech to a Los Angeles organization of construction engineers.

In a dinner meeting address in January 1960 that he titled "The Space Snafu," he once again reviewed the space shots that had made public headlines since the Sputnik event and compared the Russian achievements with those of the U.S.[22] Based on these numbers, he concluded that the Soviets had demonstrated a significant lead over the U.S. in the ability to throw heavy loads into space and to deliver them to a predetermined point in space with remarkable accuracy. This fact implied that the Soviets could deliver ICBMs to any point in the U.S. with equal efficiency. It was a cause for great concern, but not for despair. "The United States had been a late starter in the past, but had always shown a remarkable aptitude for catching up, and seizing the lead," he said.

Pickering believed that space would be no exception and there were already signs of progress in that direction, but he sensed that there was confusion in the mind of the general public about the direction that the nation's space program should take.

The public had a right to ask, "Why do we need to spend a billion dollars on a space program?" he said. The principal reason for spending a billion dollars on space, Pickering suggested, arose from the political reality of our engagement in a Cold War with the USSR. "If we are interested in having the U.S. considered a first class power in this world, then it is essential for the U.S. to have a first class space program," he said.

Now, the world faced a new situation, " . . . the arms race has become a space race . . . and the strength of the country is measured by its achievements in space and not by its armaments."

To succeed in this race, Pickering called for better public recognition of the importance of space to force the government to create a unified national space program that avoided conflict between civilian and military interests for limited resources. "Finally," he said, "the public must learn to distinguish the reality of space exploration from the fantasies of science-fiction."

Evidently, Pickering was catching the attention of officials in high places. About one month after making this speech, Pickering (along with Howard Seifert of Space Technology Laboratories and president of the ARS; George Arthur, president of the American Astronautical Society; and Guyford Stever, professor of aeronautical engineering at Massachusetts Institute of Technology) was called to testify before the Congressional House Committee on Science and Astronautics. Pickering's testimony, which essentially replicated the "Space Snafu" speech, was reported by the *New York Times*:

> . . . the space program was being hampered by confusion, indecision and increasing military domination . . . Dr. Pickering urged that, 'a truly unified national space program' embracing both military and civilian research be established under the control of [NASA]. At present, he complained, the space program lacks a clear objective and effective coordination between the civilian and military space efforts.[23]

Writing for *Aviation Week*, Ford Eastman said, "Lack of motivation, funds, or clearly defined policies were described last week as the major weaknesses impeding the U.S. space program by top space technology experts appearing before the House Committee on Science and Astronautics."[24]

In the laboratories and machine shops around JPL, work on the hardware and electronic systems for the first Rangers and Mariners made good progress in 1960. In that fact, Pickering found cause for great satisfaction, but his relations with NASA were a different matter altogether. Reflecting on this period of his career at JPL, Pickering said:

> At the beginning, I don't think that I appreciated the difference between the viewpoints of the civil service side of NASA and ourselves. As far as I was concerned, our first allegiance was to Caltech and not to the U.S. government. We were part of the Caltech community and we wanted to maintain the academic type of approach with the freedom of choice that is implicit in that [Caltech president] DuBridge supported that concept. He thought it was a proper function for a university because it was research, it was civilian, and because it was scientifically oriented.[25]

Preserving the campus-like environment that his staff found so attractive was one thing but, Pickering soon found, convincing his senior staff that the Laboratory had to now conform to the dictates of the NASA organization regardless of their personal opinions about the merits of the directions that were being passed down to them was quite another. Pickering told the senior staff:

> We have to realize that we are part of the national program . . . and the science experiments for the program should be selected on the national level and that means NASA should properly select them. . . . They accepted that, although with the egotistic [outlook] we had at the time—that we were the only experts, we got into a lot of fights with the selected scientists over their experiments.[26]

At NASA, no one was closer to the source of the problems than Homer Newell. Describing this period in *Beyond the Atmosphere* he wrote:

> As work progressed, trouble continued to brew. NASA managers came to feel that the JPL's traditional matrix organization, which might have been fine for general research and smaller projects, was totally inadequate for large-scale projects with pressing deadlines. NASA also found the Laboratory's record keeping, contract administration and supervision and reporting, inadequate. As a result, NASA began a campaign to get Pickering to tighten up the organization and to improve the administrative side of the house.[27]

NASA also took exception to the large amount of time that Pickering devoted to non-JPL matters—the American Institute of Aeronautics and Astronautics (AIAA), the International Astronautical Federation (IAF), and the International Academy of Astronautics (IAA), for example.[28]

For a time Pickering ignored NASA's strongly worded suggestion that he appoint a deputy-director to give continuous attention to the internal running of the Laboratory. Finally, however, he was forced to accede to NASA demands. Newell wrote:

> This last suggestion was especially disturbing to Pickering, who, despite NASA management's doubts about the quality of his leadership, felt keenly his role as defender of his people. The question of a deputy for the laboratory remained a bone of contention for a long time and even when one was appointed, NASA felt that Pickering did not make proper use of the position.

But NASA itself was not without blame as Newell recognized only too well:

> While the laboratory continued to insist on its independence, NASA insisted that JPL was a member of the NASA team with the same responsibilities to headquarters that other NASA centers had. Headquarters meddled too much in JPL affairs and took on too much project, as opposed to program, responsibility. Headquarters' program managers often by-passed the JPL project office and sought information or gave instructions directly to project engineering staff, or interacted directly with JPL contractors.[29]

NASA's displeasure with Pickering's heavy involvement with the IAF and other professional societies was especially painful to Pickering at this time because he believed that "it was part of his job" to represent JPL and the U.S. space program to the informed public at large. He believed that task was most effectively accomplished at the highest levels within the scientific and technical communities.

In August 1960, he had attended the 11th Congress of the International Astronautical Federation in Stockholm, Sweden. At that time, the IAF president was Academician Leonid I. Sedov of the USSR. Twenty-nine countries, including USSR, China, Germany, Japan, India, United Kingdom, France, Norway, Sweden, Denmark, Poland, and Italy sent representatives from the upper echelons of their scientific and technical establishments. The U.S. contingent included von Braun, Seifert, and Pickering who represented the ARS. An international science convention was an opportune time for a space coup as the Soviets had already demonstrated with the Sputnik affair. What would it be this time? No one knew, but a persistent air of expectancy pervaded the formal gatherings of the world's scientific and technological elite. Pickering clearly recollected with great relish what took place:

> . . . it was just after the U.S. launched the first Echo balloon.[30] The Swedes had put on a musical concert for us out at Grottingen, the Versailles of Sweden. At intermission, the people drifted outside and to their astonishment, the Echo balloon [satellite] drifted

across the sky. It was a time when satellites were few and it caused great excitement. The whole crowd came outside to look at it— except the Russians.

Immediately, Pickering was asked to brief the assembly on the purpose and technical details of the Echo satellite. There is no reason to doubt that he was fully prepared and responded to all questions with confidence. The leading French newspaper, *Le Soir*, reported the event including Pickering's details, in major headlines, "Un satellite-ballon américain a été placé sur orbite." In Pickering's view, it was another small step in the "right direction."

Casting a shadow over this whole situation was the undeniable fact that Pickering's ideas about the urgency and direction of the space program differed from those of the Eisenhower administration and those of NASA. Eisenhower doubted the value of sending a man into space, and wanted desperately to avoid a space race with the Soviets.

Within six months however, all such conflicting opinion had become moot by a succession of major political events that began with the outcome of the Presidential election in November 1960 when John F. Kennedy succeeded General Eisenhower as the nation's Chief Executive. While the nation's attention was focused on Kennedy's response to the communist threats from Cuba, yet another Russian spectacular swept unannounced and unexpected across the skies of America. It was a huge Russian space capsule called *Vostok 1* and, as if to further deride the U.S. effort in space, it carried a human payload. On 12 April 1961 the Russian Cosmonaut, Yuri Gagarin, claimed the distinction of first man to fly in space. Understandably, this event lent new urgency to Pickering's argument over the significance of "national prestige." One week later, the country's sense of pride was further tarnished by the disastrous outcome of the invasion of Cuba that resulted in the Bay of Pigs debacle.

Sensing a change in attitude at NASA, Pickering asked his senior staff to put all available effort into working up a new space plan for JPL. "The new study should take into account the primary importance of propaganda . . . etc," he directed. The new study did just that. "The primary objective," it stated boldly, "is to be first." It called for landing a man on the lunar surface in 1967, establishing a lunar base by 1969, and placing a man on Mars in 1973. While that was surely spectacular enough for propaganda, opinions among the JPL staff were varied, although generally in favor, and Pickering endorsed the proposal and passed it on to NASA.[31]

Meanwhile, the White House had set in motion the search for a national space initiative that culminated in President Kennedy's call on 25 May for the nation to commit itself to putting a man on the Moon by the end of the decade. Kennedy believed that it was essential for the U.S. to take a leading role in space: " . . . if we are to win the battle that is going on around the world between freedom and tyranny." Hearing these words from the President must have touched Pickering's pride and assurance in the value of his own beliefs. But how would NASA react? He did not have to wait long to find out. Within days, NASA issued a new space flight plan that gave national priority to a manned lunar landing, declaring that the objectives of JPL's Ranger project were now considered to be in "direct support" of Apollo. Almost overnight, the rationale for the national space program had changed. Gone was the peaceful, measured scientific approach of the past. Now it was considerations of national security surmounted by the objectives of "national prestige" that would drive the NASA programs, and Pickering felt himself vindicated.

In 1961, the American Rocket Society (ARS) claimed members of Congress, government officials, high-ranking military officers, leaders of industry, engineers, scientists, and students among its 20,000-person membership. It was the largest and most prestigious organization of its kind in the country, if not the world, and encompassed the entire missile and space business of the U.S. When the ARS spoke, people listened. In October 1960, the ARS elected William Pickering to the office of president of the society for the following year. This action obliged him to deliver the opening address at the society's annual convention in New York. The theme on this occasion was "Space Flight Report to the Nation," and Pickering planned to speak on "Space, Professional Societies, and the National Interest," themes he had delivered to other professional societies in the past.

Pickering began:

> At this moment in history, the future of our nation, indeed of the whole civilized world, depends to a large extent on the skill and ingenuity of you, the members of the ARS. Missiles for hot war, Space for cold war; these two elements of our strength are critical in determining our national posture, our standing among nations, our ability to lead the free world.

Because of its unique membership, the role of the ARS in supporting the national interest was quite clear, but Pickering questioned whether the proliferation of professional societies in recent years was "truly in the best interests of the profession [of engineering] and of the nation." Pressure to produce papers and speeches for a multiplication of professional society conferences, and the time and resources expended in attending them, could endanger the quality of the material presented. If this happened, "the whole system of technical societies will no longer be of value to the engineer and scientist and had best be abandoned," said Pickering.

Pickering commended the ARS for its efforts to achieve the highest quality in its papers and standards for membership but, he said, " . . . the quality of ARS membership may be said to be of direct interest to the national welfare. If the ARS can improve this quality, by so much will the society contribute to the national interest. There are very few societies in this position."

In 1957, the ARS had recognized its obligation to help the government by offering a recommendation for a national space flight program to President Eisenhower.[32]

Now, Pickering addressed a message directly to President Kennedy. Speaking on behalf of the ARS and its members, he said, "Mr. President, we in the American Rocket Society welcome your program for the conquest of space. We believe in it. We know it can be done. We pledge our help in every way possible."[33]

As originally conceived by NASA, and negotiated with JPL in December 1959, the Ranger program was to comprise five spacecraft arranged in two groups. Rangers 1 and 2 in the first group were intended principally to gain flight experience with the new technology required for missions to the Moon and later, to the planets. They would carry a minimal amount of sky science.[34] They would not be aimed for the Moon but boosted into a large elliptical orbit that reached part way to the Moon to prolong their flight and observing time. The second group comprising Rangers 3, 4, and 5 would be targeted to impact the Moon. They would embody more advanced technology, including a central computer brain and would carry sky science and a large array of lunar science.

These plans were in consonance with the principles that Pickering had espoused in many of his public statements: a measured approach to new technology, understand each problem before moving to the next, thoroughly understand and test new ideas before implementing them, and allow the technology to drive the schedule not vice versa.

Although they were regarded as test machines to gain flight experience, these spacecraft were in fact extremely complicated arrangements of interdependent electronic and mechanical systems. For these spacecraft to work correctly every component had to perform flawlessly—there was no room for failure. But Pickering's men were confident of their designs and were not used to being proved wrong. "We were experts, we knew how to do it," Pickering believed.[35] They resented any criticism of the efficacy of their designs and processes from the people at NASA. Last minute attempts by NASA to add more science experiments were strongly resented. In JPL's view it was "technology first" on these flights, with science as second priority. In this regard, Pickering staunchly supported his project manager in resisting NASA's demands for more science.

All five were launched between July 1961 and April 1962. In JPL's first major setback, the first four spacecraft completely failed to achieve their stated mission objectives. The reasons for the four successive failures varied, and were not all attributable to JPL, although much of the blame eventually devolved upon Pickering as Director, and provided much substance for criticism by his detractors at NASA. Of the first four Rangers, only Ranger 4 provided some cause for subdued satisfaction when its mid-course guidance system successfully directed the spacecraft into the predicted lunar impact target zone.

It was an irony of personal fate for Pickering that, in the midst of this depressing situation, the prestigious National Academy of Sciences, should announce his election to membership of that august body.

In late April 1962, the news headlines told the bizarre story:

Monday, 23: "Ranger 4 Moon Shot Racing for Target; Made in Pasadena Package Aloft." *Star-News*, Pasadena, California.

Tuesday, 24 April: "Rocket's Brain fails; Moon Shot Written Off." *Miami Herald*, Miami, Florida.

Wednesday, April 25: "Academy Cites JPL Director. Dr. William H. Pickering has been elected a member of the National Academy of Sciences it was announced yesterday in Washington . . . Dr. Pickering returned to Pasadena last night from Cape Canaveral where he had participated in the launching of Ranger 4." *Star-News*, Pasadena, California.

Thursday, April 26: "JPL Scientists Hail Feat of Hitting the Moon; Ranger 4 Strikes on the Dark Side." *Star-News*, Pasadena, California.

Although JPL could find some satisfaction in having reached the Moon with Ranger 4, the first for an American spacecraft, there could be no denying that the Soviets had already done that, and the outcome of Ranger 4 had done nothing to advance the Apollo program. A chastened Pickering could only find solace in the messages of congratulation on his election to the National Academy of Sciences that poured in from JPL, Caltech, and colleagues across the nation.

The failure of the first four Rangers provided a severe practical demonstration of the hazards of space flight, and the extraordinary precautions that were required to overcome them. Although Pickering saw this experience as making way up "the learning curve," he made sure that his senior staff also recognized this fact, and that they took strong and immediate action to incorporate the "lessons learned" into their design and test procedures for Ranger 5.

While part of the Laboratory workforce struggled with the vicissitudes of the Ranger program, a new mission which held great "space appeal" for the brilliant minds at JPL had made its appearance on the JPL task list. It was the first of the planetary initiatives that Pickering had advocated for so long, and it was called Mariner.

NASA had approved the Mariner program in mid-1961, and JPL began work on the design for its first planetary project, a mission to Venus, in the fall.[36]

The initiation of the Mariner program presented Pickering with further new problems of a type in which he had limited experience and which held little personal interest for him. Not personal but personnel—and not enough of the latter to handle the new work on Mariner in addition to the on-going Ranger project. It was hardly surprising that much of the JPL engineering staff lacked the requisite experience to adequately handle the esoteric tasks entrusted to them. The new space technologies in daily use at JPL were being invented as the work proceeded. In the infancy of the space program there was no pool of "space-experienced" engineers in industry or academia to draw upon. The Laboratory's budget had increased substantially over the years too, and its proper management, as much the responsibility of the Director as oversight of the technical issues, created a further distraction for Pickering and his senior staff.

Nevertheless, under the strong leadership of Mariner Project Manager, Jack James, work on the new project proceeded in parallel with, although somewhat behind, the Ranger project. Benefiting from the Ranger experience, James obtained approval from Pickering to set up a much more powerful project office for Mariner, with greatly enhanced authority to draw upon the best technical support available at the Laboratory to support his new project. His authority was enhanced to a large degree by the "space appeal" of the planetary mission and by the technical challenges—irresistible to JPL engineers and scientists—that went with it.

Less than a year after JPL began serious work on them, the first (of two) Mariners stood atop an Atlas-Agena booster rocket combination at Cape Canaveral ready to make NASA's first attempt to visit Earth's neighbor—Venus. It was a mighty effort but it did not succeed. Again, Pickering's hopes, dreams, and reputation were dashed by a problem not of his own making. Within the first 5 minutes of flight, the Atlas launch vehicle lost its guidance control signals and had to be destroyed when it threatened range safety boundaries. The problem was traced to a minor programming error in the computers that executed the Atlas guidance functions, and a work-around was soon developed. A month later, on 27 August 1962 the second Mariner departed for Venus—and everlasting space glory.

Finally separated from its launch vehicle and moving serenely along a trajectory that would intercept Venus in December, Mariner 2 performed flawlessly and began sending back a steady stream of new and exciting science data on the interplanetary medium. At both JPL and at NASA, the excitement and relief that followed the launch success of Mariner 2 went some way toward dissipating the sense of despondency that had arisen out of the recent string of Ranger failures. It was due to reach Venus just 108 days later.

Meanwhile, Pickering's high hopes for a fifth attempt to reach the Moon with yet another Ranger spacecraft dissipated miserably a few hours after launch. Killed by a massive, on-board power failure, the Ranger 5 became simply a piece of space junk floating interminably in orbit around the Sun. For NASA and JPL it was the "last straw."

The Ranger project, the United States' much vaunted effort to leapfrog the Soviet's demonstrated pre-eminence in space was in chaos and, to make matters worse, the chaos was highly public. All eyes turned to William Pickering, for that was—as President Harry Truman would have said—"where the buck stopped."

Both JPL and NASA were on the steep part of the learning curve now, where a great deal of learning was required to produce measurable progress, and the consequences of failure to make progress were swift and severe. Within a month, independent failure review boards convened by both JPL and NASA had delivered damming assessments of JPL's handling of the Ranger program to Pickering's desk.

Another Ranger launch should not be attempted, said the JPL Board, until the Laboratory had cleaned up its engineering design, review, testing, and management processes. Also a stronger project manager was needed: " . . . someone with a reputation for dogmatic pursuit of excellence."[37]

If Pickering was shocked by what his own people were telling him, he was devastated by the report from NASA that followed a few days later. Known as the "Kelley Report," it delivered a stinging rebuke and criticism of his management of the Laboratory.[38] At the end of it all, the report made a number of strong recommendations for change and, as a final anathema to Pickering, it proposed that NASA should exercise closer monitoring of Ranger project activity at JPL.

Pickering was embarrassed to say the least, but he had little cause for rebuttal. The study that he had endorsed in 1961 called for an even faster paced program of launches. Pickering's senior staff resented some of the allegations saying that Ranger was a high-risk project and NASA had accepted that fact right from the start. Nevertheless, Pickering had no option but to accept the criticisms and comply with the Kelley recommendations despite his personal feelings of outrage.

In subsequent negotiations it was agreed that there would be one simple objective for future Rangers: obtain a few TV images of the Moon with better resolution than pictures taken from Earth. There would be no additional scientific instruments on future Rangers and no heat sterilization. The launching of Rangers 6-9 would be postponed for as long as it took to convince NASA and JPL that there was a high chance of success for both the launch vehicle and the spacecraft. It would take them another year to learn that "a high chance" was a necessary, but not sufficient, condition of success.

Toward Venus

While the new course for Ranger was being negotiated with NASA Headquarters, Pickering endured another emotional roller coaster at JPL. For the previous 108 days Mariner 2 had been moving flawlessly along the trajectory that would intercept the orbit of Venus on 14 December. During that time it carried out an astonishing number of "firsts in deep space." It had responded to Earth commands to extend its solar panels, line up with the Sun and Earth, and stabilize its orientation in space. In mid-journey it had adjusted its trajectory to eventually pass very close to Venus. Furthermore, it had used its high-gain antenna to maintain two-way communication with Earth via JPL's new Deep Space Instrumentation Facility, which had by then expanded from the single antenna at Goldstone to include two huge 26-meter-diameter antennas, one in Woomera, Australia, and the other in Johannesburg, South Africa.[39]

During its flight to the planet it had sent back new data on interplanetary fields and particles and capped it off with a 42-minute, close-up scan of the surface during which it measured the surface temperature and determined the height and density of the cloud cover that surrounded the planet.[40] It had achieved all of its mission objectives and generated many "firsts" in space as well. It had made space history and the world was amazed. The media turned Pickering into an instant hero.

Writing for the *New York Times* under the front page headline, "Mariner Inspects Venus at Close Range; Radios Data 36,000,000 Miles to Earth," John W. Finney reported, "The United States achieved a significant 'first' in the exploration of space today by sending a Mariner spacecraft near the planet Venus to take man's first close-up observations of a planet."[41]

At a news conference in Washington, DC, James Webb, the new head of NASA, called it "a historic scientific event and outstanding first in space for our country and the free world. . . ."

But there was more than science behind NASA's obvious satisfaction with the success of the mission. Before the eyes of the world the U.S. had demonstrated, for the first time, the return of scientific data directly from the vicinity of another planet. " . . . United States had at least [last] beaten the Soviet Union in scoring a spectacular and impressive first in the space race" said the *New York Times.*

For Pickering those words suggested that his ambition "to be first in deep space" was at last becoming a reality.

Quite apart from the issues of national prestige and technological preeminence, it was also regarded as " . . . the most significant, as well as the most spectacular of the nation's scientific efforts in space thus far. . . ."[42]

Pickering must have been more than happy with the press reports that day. Mariner 2 had indeed survived the space environment and carried out a successful guidance maneuver, it had navigated 182 million miles in 109 days to a distant planet, demonstrated telemetry from deep space, and returned a substantial amount of significant science data, all matters of the deepest interest to him. Moreover, it had beaten the Soviets to it. With time of course, such engineering accomplishments would be refined and would become a standard part of JPL's remarkable repertoire of expertise in deep space technology. But then, in December 1962, such things had not been done before and Pickering and his team were the first to demonstrate their practical application where it mattered most, in deep space.

Among the first to recognize the Mariner 2 achievement at the highest level was President Kennedy. Pickering received an invitation to visit the President early in the new year to brief him on the momentous event. Pickering determined to take Jack James and Robert Parks along with him to the White House to share the honors.

A few months later, *Time* magazine featured "Physicist William Pickering" on its front cover, and a full length article titled "Voyage to the Morning Star" that praised Pickering and his JPL team for its effort and gave a very cogent account of the mission and the science results.[43]

The clear success of the first Mariner mission to Venus and the worldwide acclaim that accompanied it, together with the obvious implication that it had wrested the lead in space from the Soviets, provided a much needed lift for the flagging spirits of Pickering's hard-pressed team at JPL. But for all the hoopla, the success was a hollow one for Pickering. Coming in the midst of the Ranger debacle, the Mariner 2 success had certainly eased the palpable mood of depression that hung over JPL, but there was persistent uncertainty as to whether the new measures instigated by NASA would work out, despite NASA's added "help." Only the future would tell. The outlook from the Director's office, obscured not only by concern for the eventual outcome of Ranger, but also by apprehension over the forthcoming contract renewal negotiations with NASA in the new year, was far from clear. There, NASA would have the high ground and could use its dominant position to force Pickering to conform to its own ideas of how the Laboratory should be managed. Pickering could take small consolation in knowing that a large part of the problem was of his own making.

Pickering's angst could only have been heightened by the knowledge that NASA's demands for restructuring the Ranger organization would entail the

President Kennedy discusses the Mariner 2 mission to Venus with William Pickering and NASA officials at the White House, 17 January 1963. Left to right: Pickering, James, Webb, Parks, President Kennedy, Newell, and Cortright. James and Parks hold a model of Mariner 2 to be presented to the President (Photo: By Robert L. Knudsen, The White House).

replacement of its two principals. Both Cummings and Burke were highly popular characters at JPL and he regarded both engineers as personal friends. It was a dark day indeed, more so for its proximity to the much heralded Mariner Venus success, when he made the announcement. The bad news shocked the entire establishment. Henceforth, Robert Parks would direct both the lunar and the planetary Programs, and the new Ranger project manager would be Harris M. "Bud" Schurmeier.[44] It was the week before Christmas in 1962.

That year, Christmas Day came on a Tuesday. As was the custom, Pickering closed the Laboratory on the preceding Monday to give his people a long weekend break from the daily pressures of work. Personally, he too welcomed some respite, brief though it might be, from the "hot-seat" at JPL, and a chance to pause and "reflect on events," as he was fond of saying. This year there was much to reflect upon in both his professional and private lives.

Despite the string of failures that had dogged the Ranger program, the success of the Mariner Venus mission had brought him enormous public acclaim, mainly because it had, at least, evened the score with the Soviets, and the public desperately needed a public hero at that time. The strain of his worsening relationships with NASA Headquarters was, in a large measure, offset by the overwhelming public acceptance of his ideas and observations on the space program and his prognostications for winning the Cold War. He had been quoted in practically every reputable technical journal in the country and featured in the second volume of a book titled *Men of Space*.[45] News media across the country hung on his every word—dissent from his voiced opinions and ideas was minimal.

Requests for his appearance at scientific, engineering, and educational functions across the country were legion in 1960 to 1962. They came from professional organizations to PTA meetings, from congressional hearings to high school graduation ceremonies. He gave 20 speeches in 1960, 19 in 1961, and 11 in 1962, at locations from one side of the continent to the other.[46]

By Christmas 1962 his daughter Beth was into her second year at Cornell University in Ithaca, New York. She had demonstrated an outstanding aptitude for science and mathematics, two topics of great interest to both father and daughter. She had returned to Altadena for the Christmas period and Pickering would enjoy her lively company immensely.[47]

Now a young adult, Pickering's son Balfour was starting a career of his own in the electronics field and Pickering looked forward to spending some time with him also during the Christmas break.

In keeping with long established tradition, Altadenans were expected to decorate and illuminate the front yards of their homes during the festive season, and neighborly competition generally produced very elaborate displays of lights and animated Christmas scenes. It was a natural enterprise for the

Director of one of the nation's leading centers of technology. Having a demonstrated aptitude for all things electrical, homeowner Bill Pickering found no difficulty in changing focus from sending a Mariner spacecraft to Venus, to stringing Christmas lights across his front yard in Altadena.

That year the level of excitement in the Pickering household was well above normal. To recognize the success of the Mariner mission to Venus, the well-known Pasadena institution Tournament of Roses had elected William Pickering to the honor of Grand Marshal for the famous Rose Parade on New Year's Day in Pasadena. A spectacular annual event, watched by a national television audience of many millions and witnessed by hundreds of thousands of spectators lining the city streets, the Rose Parade was a Pasadena social milestone of the highest order that, in the closing days of 1962, involved the whole Pickering family in a frenzy of preparation and eager anticipation.

Endnotes

1 The 26-meter antenna at Goldstone, hurriedly built to support the Pioneer 3 launch in December 1958 was the first of several that eventually developed into NASA/JPL's mighty, worldwide Deep Space Network (see Mudgway, Douglas J., *Uplink-Downlink: A History of the Deep Space Network* (Washington, DC: NASA SP-4227, 2001).

2 Folder 24 in the William H. Pickering Speech Collection. Pasadena, California: Archives and Records Center, Jet Propulsion Laboratory, JPL-181, 2003.

3 Logsdon, John M., ed., *Exploring the Unknown: Selected Documents in the History of the U.S. Civil Space Program* (Washington, DC: NASA SP-4407, 2001), p. 268.

4 Koppes, Clayton R., *JPL and the American Space Program: A History of the Jet Propulsion Laboratory* (New Haven: Yale University Press, 1982).

5 Hibbs, Albert R., ed., "Exploration of the Moon, Planets, and Interplanetary Space." Pasadena, California: JPL Technical Report 30-1, April 1959.

6 Koppes, Clayton R., *JPL and the American Space Program: A History of the Jet Propulsion Laboratory* (New Haven: Yale University Press, 1982), p. 98.

7 Both the launch vehicle provided by ABMA and the upper stages provided by JPL were military items and therefore still classified, although the Pioneer probe itself was not.

8 See Folder 23 in the William H. Pickering Speech Collection. Pasadena, California: Archives and Records Center, Jet Propulsion Laboratory, JPL-181, 2003.

9 Folder 26 in the William H. Pickering Speech Collection. Pasadena, California: Archives and Records Center, Jet Propulsion Laboratory, JPL-181, 2003.

10 For a press report of this speech, see *Aviation Week and Space Technology*, 23 November 1959, p. 26.

11 Ibid, p. 338.

12 Soon after, Homer Newell, one of the scientists who visited JPL on that occasion made good on the NASA pledge by establishing the Lunar and Planetary Programs Office. Headed by NASA, but staffed by outside scientists, this office took various forms in the years to come and served NASA and the space program well in recommending projects that the agency should undertake (see Logsdon, John M., ed., *Exploring the Unknown: Selected Documents in the History of the U.S. Civil Space Program* (Washington, DC: NASA SP-4407, 2001).

13 Hall, R. Cargill, *Lunar Impact: A History of Project Ranger* (Washington, DC: NASA SP-4210, 1977), p. 23.

14 Folder 27 in the William H. Pickering Speech Collection. Pasadena, California: Archives and Records Center, Jet Propulsion Laboratory, JPL-181, 2003.

15 Folder 28 in the William H. Pickering Speech Collection. Pasadena, California: Archives and Records Center, Jet Propulsion Laboratory, JPL-181, 2003.

16 *Astronautics*, January 1960, pp. 83, 84.

17 Folder 30 in the William H. Pickering Speech Collection. Pasadena, California: Archives and Records Center, Jet Propulsion Laboratory, JPL-181, 2003.

18 Mudgway, Douglas J., Oral history interview with William H. Pickering. Pasadena, California, July 2003.

19 The role and composition of both the senior staff and the executive council, the two principal groups through which Pickering interacted with the laboratory as a whole, are sent an interoffice memo No.195, on 14 February 1963, to senior staff from W. H. Pickering. Folder 1, JPL 150, Archives and Records Center, Jet Propulsion Laboratory, Pasadena, California, 2003.

20 Mudgway, Douglas J., Oral history interview with William H. Pickering. Pasadena, California, July 2003.

21 Pickering's geniality made JPL an attractive venue for high-level visitors during his tenure as Director, as the archival records clearly show. See Folders 1-95, JPL-186, Archives and Records Center, Jet Propulsion Laboratory, Pasadena, California, 2003.

22 Folder 31 in the William H. Pickering Speech Collection. Pasadena, California: Archives and Records Center, Jet Propulsion Laboratory, JPL-181, 2003.

23 Finney, John W., "Scientist Says Indecision Curbs the U.S. effort." *New York Times*, 25 February 1960.

24 Eastman, Ford, "Space Lag Blamed on conservative Effort." *Aviation Week and Space Technology*, 29 February 1960.

25 Mudgway, Douglas J., Oral history interview with William H. Pickering. Pasadena, California, July 2003.

26 Ibid.

27 Newell, Homer E., *Beyond the Atmosphere: Early Years of Space Science* (Washington, DC: NASA SP-4211, 1980), pp. 265, 266.

28 Pickering did not accept that criticism and felt justified in using government time and resources to carry on his interests in the IAF, IAA, AIAA, AAS, IEEE, and several other professional institutions, as the archival records show. He considered those activities as "part of the Director's job in enhancing the image of NASA in general and JPL in particular." See JPL-187 and JPL-3, JPL Archives and Records, and Mudgway oral history.

29 Newell, Homer E., *Beyond the Atmosphere: Early Years of Space Science* (Washington, DC: NASA SP-4211, 1980), pp. 265, 266.

30 Echo was a large metal-coated balloon satellite intended for experiments in atmospheric drag, geodetics, and passive coast-to-coast communications.

31 Koppes, Clayton R., *JPL and the American Space Program: A History of the Jet Propulsion Laboratory* (New Haven: Yale University Press, 1982), p. 116.

32 Recommendation for a "National Space Flight Program." Astronautics, January 1958.

33 Folder 66 in the William H. Pickering Speech Collection. Pasadena, California: Archives and Records Center, Jet Propulsion Laboratory, JPL-181, 2003.

34 Sky science, as distinct from planetary science, implied investigations of the interplanetary medium, energetic particles, magnetic fields, and the like.

35 Mudgway, Douglas J., Oral history interview with William H. Pickering. Pasadena, California, July 2003.

36 Koppes, Clayton R., *JPL and the American Space Program: A History of the Jet Propulsion Laboratory* (New Haven: Yale University Press, 1982), p. 127.

37 Ibid, p. 131.

38 Hall, R. Cargill, *Lunar Impact: A History of Project Ranger* (Washington, DC: NASA SP-4210, 1977), p. 173.

39 Mudgway, Douglas J., *Uplink-Downlink: A History of the Deep Space Network* (Washington, DC: NASA SP-4227, 2001).

40 Three weeks later when Mariner 2 fell silent at a distance of 87.4 million kilometers from Earth it had established a new (communications) record for a deep space probe (Siddiqi, p. 35).

41 John W. Finney in the *New York Times*, Saturday, 15 December 1962, p. 1.

42 "The Voice from Venus," *New York Times*. Monday, 17 December 1962.

43 "Voyage to the Morning Star," *Time*. 8 March 1963.

44 Hall, R. Cargill, *Lunar Impact: A History of Project Ranger* (Washington, DC: NASA History Series, NASA SP-4210, 1977), p. 177.

45 Thomas, Shirley, *Men in Space, Vol.2* (Philadelphia: Chilton Company, 1962)..

46 Although he undoubtedly had help with the preparation of the less technical speeches, the effort entailed in presenting so much material across the nation, quite apart from the amount of transcontinental traveling that it involved, was quite amazing. Even more so, when one considers that, at the same time, he was running a major research institution. It is not difficult to understand why NASA officials expressed some concern for the well-being of the labora-tory entrusted to Pickering's care. The laboratory's record of success to that time did nothing to diminish the strength of their views. Despite the criticism however, Pickering seems to have thrived on the public recognition that attended these appearances, and reveled in the opportunities they provided to indulge his natural predilection for "teaching."

47 Beth Pickering Mezitt, Personal recollections of my father, William H. Pickering. Private correspondence with the author, May, 2003.

Chapter 6

The Steep Part of the Curve

For as long as most people could remember, the morning of New Year's day in Pasadena had always dawned fine and clear. To millions of viewers across the nation who watched the annual Rose Parade on their new color television sets that morning, the endless procession of gorgeous, flower-covered floats, spectacular marching bands and baton twirlers, traditionally-costumed equestrians and their beautiful horses, television and film stars, dignitaries, breathtakingly beautiful Princesses and the stunning Rose Queen, Pasadena's Tournament of Roses appeared almost unreal. Set against the backdrop of the towering San Gabriel Mountains, the brilliant morning sunshine and intense blue sky, it was always and altogether a made-for-television spectacular. So it was again on 1 January 1963. That year the theme for the event was "Memorable Moments."

Two weeks earlier, JPL's Mariner 2 spacecraft had reached Venus and successfully transmitted a package of scientific data back to Earth. It was a first for mankind; it was conceived and built in Pasadena, and it was indeed a "Memorable Moment." Pasadena was understandably proud of the man who made it happen, its own Pasadenan—William Pickering. As the Grand Marshal for that year, William H. Pickering would lead the Rose Parade along its three-mile route through the crowd-lined streets of the city of Pasadena.

"I didn't see much of the Parade," he recalled. "My family and I were in the lead car, but it was a great experience."

His daughter Beth remembered the excitement of choosing the new clothes, practicing the "royal wave" with her mother, early morning breakfast before the Parade began at the Tournament of Roses headquarters at the Wrigley Mansion on Orange Grove, meeting the Rose Queen and her court of Princesses, and riding with her brother Balfour in the front seat of the Grand Marshal's car. "It was a once in a lifetime experience," she recalled wistfully.

Grand Marshal William Pickering and wife Muriel wave to the crowds from the leading car of the Rose Parade, Pasadena, California, 1 January 1963 (Photo: NASA/JPL-Caltech Archives, Photo number P2298Ac).

A couple of days prior to the big event, Pickering had given an address to the directors of the Tournament of Roses at a preparade banquet.[1] It was a lengthy and insightful discussion of the impact of science on modern society and government and, unlike much of his earlier public discourse, alluded only briefly to themes of the Cold War and military versus civil control of the nation's space program. Rather than speak of the "impact" of space he suggested that we should regard space as a force that accelerates changes that are already under way like "the need to bring trained minds and informed intelligence to the solution of problems which are not only scientific but political, social, economic, and cultural as well."

Because science was heavily involved in so many of our most important social and political problems "individual citizens should have an improved understanding of what science is, how it operates and the circumstances which make it prosper," he said. Furthermore, the financial support of big science had now passed largely to state and national governmental agencies that were ultimately responsible to the whole body of citizens for their control. To exercise this power wisely, these citizens must be able to better understand the issues, costs, and consequences of the problems. "In the world of the future," he said, "many more politicians will have to learn about science and many scientists will have to learn more about the realities of the political arena."

Pickering saw encouraging signs ahead:

> Trends and changes were taking place in a more confident climate than had marked our first space venture. . . . This new emphasis will reflect less how we react to the Soviets in the field of space, but will reflect the exciting promise which space exploration offers to the whole fabric of our national life, our industry, our agriculture, our education, and to ourselves as thinking beings.

And, although the world could not yet see it, Pickering's "new emphasis" did indeed come to pass, as we shall see.

The Pickering family spent the afternoon of that day as principal guests of the Tournament of Roses at the famed Pasadena Rose Bowl college football game, where championship teams from the Pacific and Western conferences met each New Year's Day to compete for Rose Bowl honors. Pickering recalled enjoying the game much more than he had enjoyed the parade.

That night Pickering grabbed a selection of the Mariner 2 images of Venus from the science teams at JPL and took the red-eye flight to Washington, DC. Along with a select group of NASA officials, he was due at the White House the following day to brief President John Kennedy on the initial findings of the world's first close-up view of another planet. Pickering no doubt assured the President that the blurry images of the darkly shrouded planet marked a significant forward step in closing the gap on the Soviet lead in space, a topic of great concern to the President at that time.

Back at JPL a few days later, Pickering turned to the pressing tasks at hand for 1963. Under the highly focused direction of Bud Schurmeier, a work force that reached as high as 900 engineers, technicians, and scientists reworked the original Ranger designs to meet the new guidelines for success of Ranger 6. The mission objective was very clear—a few television pictures of the Moon at better resolution than images taken from Earth—and everything was to be directed toward that end. Confident that Schurmeier brought the necessary motivation, knowledge, and experience to the task, Pickering regarded it as his responsibility to see that Schurmeier was provided with the necessary resources to carry it out and was shielded, as far as possible, from distracting demands for attention from NASA program management.

While this important work continued at JPL, Pickering rode the wave of public acclaim that followed closely upon the Mariner Venus success. What would eventually become an ever-increasing stream of civil honors began when the Association of Engineering Societies elected William H. Pickering as the "Engineer of the Year for 1962." The Association presented its George Washington Medal for Engineering Achievement at a sumptuous banquet at the Beverly Hilton Hotel in Los Angeles in February. In accepting the award, Pickering spoke eloquently of the need for unity of interest between engineering and science and the avoidance of debilitating competition for private and public funds. "There were lessons to be learned from George Washington's admonishment to 'form friendships with all, but entangling alliances with none'," he concluded.[2] Over the past several years, Pickering had been the recipient of numerous awards from the U.S. Army and Air Force for his work on military programs, but the George Washington Medal represented the first, of many, awards from the civil sector.[3]

In April he gave a lecture on "Mariner 2—First Spacecraft to Venus" at Columbia University in New York, and a week later a lecture on "Man at the Threshold of Space" at UCLA in Los Angeles. Two speeches in May, one at a conference on "The Peaceful Uses of Space," at the University of Illinois, the other on "Business Publications" at a business management conference in Del Monte, were followed by "Some Thoughts for a Graduating Class" at the Polytechnic school in Pasadena in June and "Some Thoughts on Guidance Systems" to the American Institute of Aeronautics and Astronautics in Boston in August and "Frontiers in Space Instrumentation" to the Instrument Society of America in Chicago in September. All illustrated the wide diversity of his interests and the rising demand and popularity of his public appearances, which showed no sign of diminishing.[4]

In November 1963, the newly-formed American Institute of Aeronautics and Astronautics (AIAA) held its inaugural meeting in Pittsburgh, Pennsylvania. Formed by the merging of the American Rocket Society (ARS) and the Institute of Aerospace Science (IAS), two great aerospace societies of the day, the AIAA owed much to the energies of William H. Pickering. As Pickering recalled many years later, in his inimitable way:

> I was president of the ARS and Gene Root of Lockheed was president of the IAF. We got together with a few other people and decided that the two engineering organizations really ought to join forces, because the aeronautics people were getting interested in rocketry and the rocket people were getting interested in flying things. The next problem was who should run it, and they decided I should. So I became the first president of the AIAA.[5]

The official merger went into effect in February of 1963.

Nearing the end of the year, in November Pickering delivered the first presidential address to the AIAA with the title "Exploration of Deep Space."[6] Pickering observed that the role of science and technology in the civilization of the future would continue to grow, making it essential that the young people of the day be properly prepared to take responsible parts in the new era. Nevertheless, it was important to keep a proper balance and not try to make everybody into a scientist while recognizing that "a knowledge of science is going to be the mark of the well educated man of the future." He then delivered a lengthy but masterful lecture, illustrated with slides, that covered the whole gamut of the technology of space exploration as it was known at that time. The motions and relative distances of the Sun and planets, spacecraft, launch vehicles, navigation, guidance systems, communications systems, scientific instruments, rationale for space exploration, status of Soviet space programs, and, finally, examples from his recent personal experience of the successes and failures of the U.S. space program were all included. Only Pickering could have given a presentation of that scope and depth and enhanced it with the credibility of personal experience.

It was a dazzling performance that must have held his audience spellbound, for everything that he told them was utterly beyond the experience of the vast majority of the engineering professions of the time.

He ended on a note of high anticipation for the success of Ranger 6, which, he implied, embodied all that had been learned so painfully from the experiences of past launches.[7]

Despite outward appearances to the contrary, a dark cloud of dissension had gradually gathered about JPL during the past couple of years. The problem concerned management relationships between JPL, Caltech, and NASA.

Pickering had enjoyed a large measure of independence from NASA-imposed controls through 1961. It was, after all, hard to argue with success and the NASA bureaucracy was forced to accept JPL as a brilliant, but arrogant, outsider to the NASA family of civil service Field Centers. Pickering did whatever was necessary to achieve JPL's objectives as he perceived them, and he ran its organization, management, and financial accounting practices in a loose arrangement more akin to a university administration than that of a civil service institution. Pickering believed this type of working environment attracted the best people and that they were the key to JPL's early successes. Alluding to NASA's control over JPL programs, Pickering once remarked wistfully, "Why don't they just give us the money and go away?"[8]

All of that changed in 1962 as the string of Ranger failures began to mount. "The academic, relaxed atmosphere that pervades the JPL Campus . . . did not encourage quick responses and strong team efforts on project-oriented tasks," complained one NASA official.[9] Pickering's indirect management style came under criticism too, as did the excessive time he spent away from the Laboratory on public appearances and in non-NASA matters, particularly the AIAA.[10] NASA also found deficiencies in JPL's business procedures, property accountability, contracting administration, and security arrangements.

What was needed, said NASA in January 1963, was a strong general manager to "oversee day-to day operations." Sensing that both his authority and his ability to run the Laboratory were being challenged by an entity for which he held little respect, Pickering refused to accede to NASA's request. As Pickering perceived it, the business management of JPL was not the problem—NASA's interference and oversight of JPL's internal affairs was the problem.

Pickering also faced internal problems with his highly motivated technical staff. It was difficult to persuade his senior staff to accept the NASA approach to the various technical issues about which they held such divergent opinions. But with a long string of Ranger failures on the record they were in no position to argue. Recalled Pickering:

> If Ranger had been a success right from the start, I would have had a much more difficult time keeping those guys in line. But we were forced to admit that we were having troubles. Even though we were the best in the world, there were people from outside who could tell us what we were doing wrong—that was hard to accept.[11]

These issues came to a head during the negotiations for renewal of the NASA-Caltech contract in the latter part of 1963. NASA officials, particularly Administrator James Webb, had long been dissatisfied with the so-called "management" fee—the fee that Caltech charged NASA for its oversight of JPL. Together with the lack of responsiveness and poor performance of JPL, compounded by Caltech's disappointing involvement in research support for JPL's scientific programs, the issue of management fee gave NASA good cause to press Caltech for changes to the existing arrangements. The demands included the appointment of a general manager for JPL. Some in NASA felt so strongly about the issue that they talked of canceling the contract entirely, or even replacing Pickering as Director.[12]

Caltech, on the other hand, wished to increase the fee, arguing that an increase was justified by the increase in JPL's budget. At the same time, DuBridge was acutely aware that Caltech could not afford to lose the lucrative NASA contract that had, by then, become a substantial part of the institution's financial resources.

Meanwhile Pickering, apprehensive about any further NASA encroachment on his domain at JPL, continued to resolutely resist any change, particularly in the matter of appointing a business manager.

In the outcome of the acrimonious negotiations that followed, it quickly became apparent that neither side could afford to lose the other. Eventually the two sides hammered out a compromise by which Caltech agreed to forego its demands for a fee increase in exchange for a restructuring of the basis on which NASA estimated the fee. Pickering also agreed to an annual evaluation of JPL's management, technical, and schedule performance.[13]

Satisfied with the outcome of the contract negotiations, and pacified with the promise of a new business manager for JPL, NASA deferred signing the agreement until it could evaluate Pickering's response to its provisions.

For Pickering, the aftermath of the 1963 contract negotiations was less acceptable. For all intents and purposes, the Laboratory had now become a government-furnished facility subject to the authority of NASA. Still worse, it would be subject to an annual performance review by NASA program managers, an embarrassment not imposed upon other NASA Centers.

Pickering responded with a token gesture. In December he appointed his former student Brian Sparks to the position of Assistant Director for Technical Divisions for the Laboratory. It was a transparent foil that fell far short of satisfying NASA's demand for a fully empowered deputy that would manage the Laboratory's business operations.[14]

Pickering seemed oblivious to the adverse reaction that this move caused at NASA. At the end of 1963, his attention was more likely focused on the events taking place in the clean rooms of JPL's spacecraft assembly area where sterile-clothed engineers were in the final phases of the preflight qualification tests of the lunar spacecraft Ranger 6. Success was on everyone's mind. Nothing would help JPL's tarnished image at NASA so much as a successful Ranger flight to the Moon culminated by a handful of close-up images of the lunar surface.

The exigencies of the contract negotiations and the strained relations with NASA had little effect on the general feeling of optimism that pervaded the offices, conference rooms, laboratories, and machine shops around the Laboratory in January 1964. They were, as Pickering intended "shielded from all that distraction." Focused on the job in hand, engineers reworked Ranger 6 to embody all of the changes and additions that could conceivably contribute to ensuring a successful mission. Taking advantage of the additional weightlifting capacity of the Atlas launch vehicle, spacecraft designers added redundant units for critical spacecraft elements such as the attitude-control system, the radio transponder, and the solar-powered battery charging system. It was a tradeoff of spacecraft weight versus improved reliability.

Most, but not all, of the Laboratory's personnel were preoccupied with Ranger 6 that January. A small number were involved in making final arrangements for the Coronation Ball, a very popular annual social function which, that year, was planned for late January at the Moulin Rouge ballroom in Hollywood. Known as "The Queen of Outer Space Ball," tradition demanded that the Director and his lady dignify the function with their presence, and that the Director perform the culminating function of crowning the chosen "Queen of Outer Space." It was an obligation that Pickering willingly accepted and had performed enthusiastically throughout his tenure as Director. However, as the Ranger 6 launch schedule slipped into late January it raised the possibility that someone other than the Director might have to crown the "Queen" that year.

By the end of January, Ranger 6 had been trucked to the Florida launch site, mated with the awesome Atlas-Agena launch vehicle, and successfully passed the rigors of its final prelaunch tests. Encapsulated in its protective shroud at the very tip of the two gleaming rockets that would hurl it to the moon, Ranger 6 waited as the final seconds of the countdown elapsed. Then, with a mighty shove from its Atlas launch vehicle and a powerful kick from its second-stage Agena, Ranger 6 finally shrugged off Earth's gravity and headed out toward the Moon. The date was 30 January 1964. The spacecraft was due to arrive on the lunar surface 68 hours later.

Except for an unexpected glitch shortly after liftoff, when the television system suddenly came on and just as suddenly switched off, the launch was perfect. Puzzled by the unexpected television incident and unable to come up with an explanation for its occurrence, or to see anything wrong on their telemetry monitoring channels, the spacecraft engineers elected to let the mission continue. It was the logical thing to do—or so it seemed at the time. Right on time, Ranger 6 executed a perfect mid-course maneuver that adjusted its trajectory to target the chosen point of impact on the lunar surface.

All seemed well, and to the engineers and officials anxiously waiting at JPL and NASA Headquarters, success seemed within reach. The spacecraft could not do other than continue on to impact the lunar surface, assuredly transmitting the long-awaited pictures as it did so.

During the long wait from launch, a crowd of newsmen and television crews had assembled in JPL's new Von Kármán auditorium, where an official commentator, television monitors, and frequent press conferences kept them informed of the progress of the mission. As hopes for success mounted around the Laboratory, Pickering told the newsmen, "I am cautiously optimistic."[15] His caution turned out to be well-founded.

Shortly after midnight on 2 February, in full view of a packed auditorium, the Ranger 6 mission drew to a swift and terrible conclusion. Eighteen minutes before impact, the TV cameras began the warm-up sequence as scheduled. The tension in the Von Kármán auditorium became palpable as the audience agonized through the final 5 minutes before the first television pictures should appear. But none came. JPL announcer Walt Downhower counted down the dreadful minutes to impact as Ranger 6 barreled in toward the lunar surface at over 4,500 miles per hour, blind and beyond help. He reported impact at 1:24 a.m. Pacific Time.

It was over. Ranger 6 too, had failed. After a perfect flight, the unthinkable had happened—the television cameras had failed to turn on. Why? Nobody knew nor, at the time, could they even conjecture.

Up in the visitor's gallery of the new Space Flight Operations Facility, Pickering had gathered with Homer Newell and several other important guests to listen to the voice commentary from Goldstone tracking station. At the fateful announcement, "still no video—impact," both men were momentarily struck speechless. In any case, words were unnecessary. Each knew what the other was thinking and what this result could portend for JPL and for Pickering in particular. Pickering recalled the dramatic moment:

> Then we had to go over to the Von Kármán, and the place was full of the press, and tell them it had been a failure. It was made all the worse because everyone was so optimistic, the flight had gone so smoothly, right on target up to that point. In fact, Bud

Schurmeier had laid in a good supply of champagne to celebrate, but instead we had to go through this disaster. I never want to go through something like that again—ever. It was probably the lowest point in the Lab history. I was very concerned about how the Laboratory personnel would regard it after all the other failures. . . .[16]

He was about to find out. As it turned out, William Pickering and Muriel did make it to the Queen of Outer Space Ball, which took place a few days later. He entered the packed ballroom with a heavy heart, uncertain as to how the crowd would react. But, to his utter amazement, he was greeted with standing applause and wild cheers of encouragement. "That show of confidence made me feel very good," he said, "the people realized we had a problem but they were going to solve it and not give up."[17] And later in the evening he crowned the Queen, just as he had done many times before. But the sparkle and excitement of the evening gave no hint of what lay ahead for the Laboratory.

The immediate fallout from the Ranger 6 disaster produced two committees of inquiry: one convened by JPL and the other by NASA. Both review boards determined that high-voltage circuits that powered the television camera and transmitter had been destroyed by electrical arcing as the launch vehicle passed through Earth's upper atmosphere. That was the probable cause of the TV "glitch" observed shortly after lift-off. There, the agreement ceased. The JPL review board believed the basic design was sound and it suggested ways in which the problem could be fixed.[18] The NASA review board thought differently—very differently.

Headed by Earl Hilburn, a harsh critic of Pickering's administration from the earlier contract negotiations, the NASA review board brought an aggressive attitude to the investigation. It challenged Pickering's long-held concepts for standards of excellence in design and testing and pointed to what it perceived as major deficiencies in JPL's prelaunch test procedures. The suggestions of incompetence that were implied in the report's harsh criticism included the television system contractor, JPL, and even the NASA Office of Space Science and Applications.[19]

Inevitably, a Congressional investigation would follow with William Pickering as its prime witness.

The Congressional hearings began on 17 April in Washington, DC, under the chairmanship of Representative Joseph Karth. Prior to the hearings, Karth made several visits to JPL to familiarize himself with the space program and JPL's role in it. On each occasion the Karth party was hosted by Victoria Melikan, Pickering's newly arrived manager for public affairs.[20] As a consequence of these visits, JPL officials came to respect Karth's show of interest in the subject of his investigation, and to regard his judgment as an important indicator of public wisdom and one which they could not afford to dismiss lightly.

Director William Pickering presents roses to JPL's Queen of Outer Space at the Coronation Ball, Moulin Rouge, Hollywood: February 1964 (Photo: NASA/JPL-Caltech Archives, Photo number P32568B).

Melikan's handling of the Karth Committee visits to JPL helped immeasurably to enhance JPL's image in the hearings that followed, and set a precedent on which Pickering came to depend for his public relations for the rest of his career at JPL.

During the Congressional hearings, Pickering remained unapologetic. While conceding that there had been management problems at the Laboratory and valid criticism of its business practices, he asserted that ultimately, JPL had always responded to NASA's technical direction, and he argued that the freedom associated with the university type of atmosphere he had created at JPL was conducive to the unique type of work carried out by the Laboratory.[21]

In his typical understated style Pickering recalled, "It was the first time I had been called up before a committee in just that way, and also of course put in a defensive position. It was quite an experience listening to the [arguments] back and forth across the committee."[22]

In the end, the subcommittee found shortcomings in NASA's oversight of JPL and recommended that NASA should exercise closer control over Laboratory activities, particularly its Ranger program. It drew attention to Pickering's reluctance to accept direction from NASA, calling it "embarrassing unwillingness," and recommended that NASA install a general manager as a deputy to the Director.[23]

Undeterred by the controversy and criticism that swirled about them in the aftermath of the Ranger 6 debacle, Schurmeier and his team, strongly supported by a strengthened team from Radio Corporation of America (RCA), set about readying yet another Ranger for yet another attempt to reach the lunar surface with all systems "go for impact." Ranger 7 would embody all that had been learned, deduced, analyzed, and surmised from the Ranger 6 debacle. As Cargill Hall wrote, "All of Ranger's participants very clearly understood that personnel changes were likely in Pasadena and in Washington should Ranger 7 also fail and the Hilburn Board's contentions be proved accurate. The tension was correspondingly magnified and the pressure to succeed now was unbelievable."[24] By mid-June, the NASA "Buy-Off" committee had reviewed all of the test records and reports and determined that Ranger 7 and its modified television subsystem met, or even exceeded, the established test criteria. With NASA approval in hand, the JPL team moved Ranger 7 to Cape Kennedy for mating and a final round of testing with its Atlas-Agena launch vehicle.

On 28 July, the scene was once again set for the unfolding of another Ranger space drama, but this time the stakes were higher than they had ever been, and for no one were they higher than for William Pickering. His professional career and his reputation now rested on the outcome of the Ranger 7 mission. As the mission played out over the next 68 hours, everything fell into place: launch, separation, solar panel extension, high-gain antenna deployment radio signal strength, midcourse-maneuver, camera warm-up, and a myriad of telemetry measurements all sequenced perfectly.

Newsmen from around the world had begun gathering at the Von Kármán auditorium days earlier and, together with several hundred sleepless and tensed JPL employees, listened to Ranger 7's final moments of glory as the JPL announcer relayed events from Goldstone. Cargill Hall recorded the countdown:

> Five minutes from impact . . . video signals still continue excellent . . . everything is GO as it has been since launch. . . . Three minutes . . . no interruption, no trouble. . . . Two minutes all systems operating . . . pictures being received at Goldstone. . . . One minute to impact . . . excellent . . . signals to the end . . . IMPACT!

Abruptly, the hum from Ranger's distant radio telemetry signal ceased, only the low hiss of electronic noise remained on the loudspeakers. Cheers, and many tears, erupted throughout the packed auditorium. To most of those present, including this author, the event seemed too surreal, too Hollywood movie-like to be real. The unlikely, gleaming machine that most of us had worked on, or seen many times in the spacecraft assembly area at JPL, was actually on the Moon, and we had pictures to prove it. It hardly seemed possible.

Within the hour, President Johnson called to congratulate Newell and Pickering, NASA and JPL, and its industrial contractors. He invited Newell and Pickering to the White House to brief him on the Ranger 7 findings the following day. Joined now by Project Manager Bud Schurmeier, Pickering and Newell made their way over to the Von Kármán auditorium to be greeted with a standing ovation. "How different from last time," mused Pickering. "This is JPL's day and truly an historic occasion," observed Homer Newell. "We have had our troubles," Pickering reflected ruefully, "but this is an exciting day." When a newsman asked how he viewed the Laboratory's future after the success of Ranger 7 Pickering promptly replied, "I think it has just improved." When the laughter and applause subsided, he gave credit for Ranger's success to Bud Schurmeier and the Ranger teams at JPL, NASA, and in industry. It was all very appropriate.

The following morning, Saturday, Pickering and Newell flew to Washington with a selection of the Ranger 7 pictures to brief the President.

Evincing more interest in the geopolitical implications of Ranger's success than in its scientific import, the President commented: "We know this morning that the United States has achieved fully the leadership we have sought for free men. This is a battle for real existence in the world isn't it—for survival?" That remark would have resonated with William Pickering—it echoed the main thrust of much of his public advocacy over the past several years.

The Ranger images, over 4,000 of them, astounded scientists and public alike with their extraordinary clarity, revealing thousands of craters of varying size from meters to hundreds of meters in diameter, far more than anyone expected. Scientists exulted in the new data and what its subsequent analysis might tell them about the formation of the Moon, and the nature of its surface. Apollo mission designers were pleased to discover that the lunar surface was smoother, and therefore less threatening for a lunar landing, than they had expected.[25]

The reaction to the Ranger success was no less jubilant in Washington than in Pasadena. NASA officials, reporters, and a large Congressional delegation that had listened to the proceedings in the auditorium at NASA Headquarters were overjoyed at the outcome and congratulations flowed freely. Of particular note, House Space Committee Chairman George Miller was prompted to declare that Ranger 7 " . . . puts us well ahead of the Soviets in the exploration of space." Referring to the recent Karth Committee investigation, he added, "I want to make it crystal clear that the Jet Propulsion Laboratory is doing a splendid job."[26]

Cargill Hall wrote, "On newsstands at the airports and across the land, papers acclaimed the Ranger 7 and its lunar pictures in superlatives—on the front page and in editorials. From the Seattle *Post Intelligencer* to the *Miami Herald*, or the Boston globe to the San Diego *Union*, the praise was unanimous. Overseas the foreign press responded in similar vein, seeming to agree, that the U.S. had at last forged ahead of the Soviet Union in space exploration. Even the Soviet press accorded the flight modest plaudits, though pointing out that the USSR had photographed the moon five years before. The glowing accounts frequently heralded Ranger 7 as the greatest advance in space research since Galileo had trained his telescope on the heavens—it was heady stuff."[27]

The *New York Times* recognized the Ranger 7 achievement with blazing front-page headlines that hinted at its implications for the Apollo manned missions to follow, rather than its effect on the U.S. position in the space race with the Soviets.

New York Times writer Richard Witkin reported that the " . . . details of the lunar region were seen as one thousand times clearer than before," and hailed the feat as " . . . a leap in knowledge." As an indication of the national significance of the event, the *New York Times* reported the full text of the post-flight news conference at JPL, devoting altogether at least four full pages in its Saturday edition to Ranger 7. Taking advantage of an opportunity to extol the virtues of its television technology, RCA took out a magnificent full page advertisement in that same edition emphasizing its association with the successful Ranger 7 mission.

First image of the Moon taken by a United States spacecraft: Ranger 7 took this image on 31 July 1964 at 13:09 UT about 17 minutes before impacting the lunar surface. The area photographed covers about 360 km from top to bottom. The large crater at center right is the 108 km diameter Alphonsus. The Ranger 7 impact site is off the frame to the left of the upper left corner (Photo: NSSDC image ra7 B0001; also available online at *http://nssdc. gsfc.nasa.gov/planetary/lunar/ranger*).

The following day, Sunday, 2 August, the *New York Times* again devoted its leading front page articles to Ranger 7, including close-up pictures of the lunar surface, and a picture of William Pickering briefing the President on the Ranger 7 photographs. Reporting on this visit, *Times* writer Tom Wicker said that scientists had told the President that the Ranger 7 pictures " . . . had demonstrated that selected lunar areas were suitable for manned landings." He also reported that the President had " . . . turned the occasion into a resounding endorsement of the moon-landing project and a justification of the American space effort."

As he returned to Pasadena, Pickering would have had good reason to feel vindicated over his leadership of JPL's space program and perhaps excused for his impatience with those at NASA who thought otherwise. It was easy to be critical with failure but hard to argue with success.

The Congressional Subcommittee's recommendation that "JPL should appoint a strong general manager as deputy for the Director" was timely, but somewhat superfluous for, by then, Caltech had already initiated action to find a suitable man for the job, despite the objections of William Pickering.

The successful candidate was Alvin R. Luedecke, a former Major General in the U.S. Air Force, and retiring general manager of the Atomic Energy Commission. To all appearances, Luedecke was the perfect man for the job. NASA was delighted and Caltech was greatly relieved to have, as it thought, cleared up the matter. But Pickering was dismayed. With wide-ranging responsibility for day-to-day management of Laboratory resources and authority for the direction of its financial, technical, and administrative activities, Luedecke represented that very situation that Pickering had resisted for so long: a palpable NASA presence in his Laboratory and a powerful dissenting voice in its operation. Luedecke took up office in August 1964 right after the Ranger 7 euphoria had subsided, and the struggle between Pickering and Luedecke began.[28]

Pickering would recall his concerns at the time:

> In view of all the criticism that had gone on about me I was in no position to resist their choice [for Deputy Director]. I agreed in principle to changing the flavor of the Laboratory with a deputy, but I wanted to do it on my terms rather than on his terms. Right from the beginning it was clear that we had different philosophies about things. He wanted to bypass me when he could, and go directly to NASA. [Obviously] we could not work together.[28]

For Pickering, the arrival of a general manager on the Laboratory staff was a bitter letdown after the sense of elation that followed Ranger 7. However, hard though it was for him to accept it personally, the reality was that the character of the Laboratory had changed with its increasing status as a government-sponsored institution. Now accountable for an annual expenditure of hundreds of millions of government dollars, JPL could no longer expect to enjoy its former freedom, no matter how desirable that may have been. But that was not the end of the matter, as subsequent events would soon show.

Pickering in Public (1963–1964)

If William Pickering's productivity was slowed down, or in any other way adversely affected by the dissension that swirled about him in 1964, it was certainly not apparent in the quality, or quantity, of his public discourse. From thirteen in 1963, the number of his public speeches increased to twenty-one in 1964. Together with the publication of five technical papers, this effort made 1964 one of his most productive years. When reminded of this fact in later years, and questioned about the time away from the Laboratory that the transcontinental and international speech-making tours entailed, he was somewhat taken aback. "I never realized it was that many," he said, allowing that he had received much help in preparing his speeches from Harold Wheelock, a JPL writer who had " . . . a good turn of phrase."[29] The two speeches that follow are illustrative of Pickering's genius for choosing a topic to match the dominant interest of his audiences, the economics of space for the bankers and spacecraft guidance for the engineers, and for conveying an equal sense of credibility, depth of knowledge, and enthusiasm for both.

At a time when his future, and that of his Laboratory, balanced on the fate of Ranger 7 then in its final stages of prelaunch testing under the baleful eye of NASA monitors back in the Pasadena, it was somewhat ironic that Pickering was discussing the topic "Space—Boon or Boondoggle?"[30] with the California Investment Bankers Association in Santa Barbara. "The United States space venture, like any momentous objective, is always controversial," he said. "The argument was not about whether there should be a space program, but rather about how much of the nation's limited resources should be devoted to it. The present debate was . . . whether or not the present rate of development of the space program was warranted."

Pickering saw the Kennedy-inspired initiative to place a man on the Moon by 1970 as a goal, rather than a "fiat," that would never be achieved by leaders of the program who honestly doubted the feasibility of the goal itself. "In any such operation, with a time limit for its completion . . . the major element of success is largely found in the enthusiasm and drive of key personnel which expands like contagion and inspires the team to do the impossible," he said.

Pickering felt that another necessary prerequisite to achieving the goal by 1970 would be gathering the necessary physical data (on the lunar surface conditions) to ensure the success of a manned landing. This could be done with instrumented robot spacecraft, he said.

While many prominent leaders had questioned the diversion of engineering and scientific talent to the space program at the expense of other fields of research, Pickering quoted NASA data that showed "only six percent of the national manpower pool in science and engineering was devoted to NASA contracts with private industry, plus an additional one percent in government laboratories." This was not an "overwhelming drain" for a program of such substantial value to the nation.

Pickering observed that, while arguments in the economic, scientific, and military areas were far from conclusive, it was in the human and psychological area that "the real catalyst is to be found which precipitated the program."

Given that the space race is economically possible and scientifically feasible, "Is the space race a valid psychological weapon in the battle for world leadership between the USA and the USSR?" he asked. "Armed might, while an essential element of United States world prestige and leadership, is not sufficient to ensure preeminence in the world community," he argued. "In the modern complex society, nothing adds more prestige than technological preeminence."

Pointing to the world attitude after the "Sputnik affair" as an example, he continued, "Technological supremacy is a key tool in winning the admiration and the minds of men for our system against its communist and collectivist competitors." Unlike the space race, few other projects would be "immediately understood, by the poorest coolie in the rice paddies of China or the bush dwellers of darkest Africa," he said.

"The discovery and investigation of new horizons of knowledge has ever been an insatiable yearning of mankind," he observed. Whether or not the international prestige of the U.S. warranted the cost of the space program, there were "numerous other blessings which flowed into our economy, into our scientific technology, and into our human ego, which together or by themselves, made the program a boon and not a boondoggle."

Within a few weeks of General Luedecke's arrival at JPL as General Manager, Pickering took off on a lengthy international lecture tour that included Poland, South Africa, and Spain. It began at the annual congress of the International Astronautics Federation in Warsaw, Poland. There, in company with Bud Schurmeier, he generated enormous enthusiasm for the U.S. space program when he showed a movie of the Ranger 7 lunar mission.

In South Africa a week later, he embarked on a lecture tour of several well-established universities at the invitation of the South African Institute of Electrical Engineers. The lecture in Johannesburg was the centerpiece of his South African tour. "Guidance for Interplanetary Spacecraft" was a speech of the kind where William Pickering had few equals—an esoteric, technical subject in a field where his unique experience commanded profound respect and his technical achievements and aspirations for the future of space were unchallenged and admired.[31] It was delivered to a professional audience in a university lecture style with many slides to illustrate salient points.

Travel through space he said:

> . . . introduces some fundamental new concepts. By contra-distinction with the airplane, the spacecraft does not require the application of a continuous forward thrust, nor is it subject to external disturbing forces. For most of its journey it travels freely, without drag, falling through the gravitational field . . . in free fall, the path is exactly determined by the shape of the gravitational field and the initial position and velocity of the spacecraft." A precise knowledge of the so-called 'initial conditions' allowed spacecraft navigators to determine the position of a spacecraft 'for all future time.'

Interplanetary spaceflight, Pickering explained, involved two major and quite distinct, phases. In the first phase the spacecraft plus booster rocket is controlled along a predetermined path to establish an initial position and velocity. In the following phase, the spacecraft is controlled in attitude, and minor adjustments are made to its velocity to correct for initial guidance errors that occurred during the powered flight period. He went on to describe how guidance and control was accomplished in each of these distinct periods and the methods used to establish reference directions for each: gyros or inertial platforms for rockets and celestial references of Earth, the Sun, and stars for spacecraft.

The problem of calculating the path of a spacecraft traveling to another planet was complicated by the fact that the path was determined by the influence of three successive gravitational fields: first that of Earth, then by that of the Sun, and finally by the gravitational field of the target planet. "The complete journey is therefore best described in three different coordinate systems which must be carefully matched to achieve the desired target accuracy. Without a modern high speed digital computer, this would be impossible to solve in any reasonable time," he said.

Using the recent Ranger flight to the Moon and the Mariner flight to Venus as examples, he showed how the separate orbital motions of Earth, Moon, and Venus and the arrival conditions required by the science experiments limited the opportunities for launch to very tightly constrained "windows of duration" in both time and calendar date.

After a spacecraft had been launched successfully and injected on to a path close to the nominal trajectory, it became necessary to make a precise determination of that trajectory in order to know if the targeting requirements would be met. For that purpose a high precision, ground-based radio tracking station such as the one near Johannesburg was required.[32]

Concluding, Dr. Pickering said that with the potential solution to the problem of guidance for interplanetary spacecraft demonstrated by Ranger and Mariner " . . . it is now possible to explore the solar system. Landings on the

Moon close to desired targets and flights passing near other planets can be made, and spacecraft can be sent into orbit around the moon or planets." It would not be long before William Pickering made good on his prescient observation.

In September 1964, the Italian city of Genoa, acting through the office of the Mayor and upon the recommendation of the International Institute for Communications, awarded the Medaglia Colombiana 1964 (1964 Columbus Medal) to William Hayward Pickering, with the following citation:

> For high executive skills, for the decisive contribution made to research and experiments on space probes for interplanetary exploration, and for the success achieved with the Ranger VII, not so much from the standpoint of ballistics of the future as because of the scientific value of the pictures taken (as many as 3900) in the last minutes of the mission, which constitute in the whole, a clear demonstration of the manner in which electronics and the information theory, smooth the way for the explorers of cosmic space.

The Mayor noted that "Professor Pickering was in Genoa in October 1959 as speaker before the 7th International Communications Conference, where he submitted a most interesting paper on 'Communications with Lunar Satellites' in which he anticipated the underlying causes for the success of Ranger 7."

Because his speaking schedule in South Africa prevented him from attending the Columbus Day celebration in Genoa to receive the medal, Pickering asked U.S. Ambassador Frederick Reinhardt to accept the award in his stead.

So it was done. In a brief acceptance speech Ambassador Reinhardt pointed out that Professor Pickering "was particularly anxious for his Italian colleagues to know that he wished to share the honors bestowed upon him by the city of Genoa with the other scientists and engineers who made possible the success of Ranger 7."

With obligations to his South African hosts completed in mid-October, Pickering returned to California. He was anxious to catch up with the progress of the Ranger 8 lunar spacecraft at JPL but more importantly, for the moment, to support Jack James and his Mariner team as it worked through the final prelaunch test sequences for Mariners 3 and 4 at Cape Kennedy.[33]

Toward Mars

The first mission to fly by Mars presented its designers with a number of significant technical challenges that took them well beyond those associated with lunar missions. Many of the problems were those very guidance and control issues that Pickering had discussed on his South African tour.

Mars was about three times further from Earth than Venus, which meant that the tracking stations would see only about one tenth of the signal power from Mariners 3 and 4 than they had seen from Mariner 2 at Venus. Because the spacecraft distance from the Sun would be greater too, the Mars spacecraft would require larger solar panels. Nor could these spacecraft use Earth as a reference point for their attitude stabilization system. To the Mars-bound spacecraft, Earth would appear only as a faint crescent-shaped object against a bright star background at those times when it was not obliterated by the light from the Sun. Canopus would be the celestial point of reference for Mariners 3 and 4, but how would they find that particular star amongst the thousands of other bright objects in the celestial sphere?

Spacecraft designers would have to balance the increased weight of the video cameras and their articulated mounting platform together with the other science instruments, against the payload weight limitation imposed by the launch vehicle's ability to throw the heavy spacecraft the extra distance to Mars.

And then too the new spacecraft would take almost nine months to reach Mars compared with a mere three months taken by Mariner 2 to reach Venus. That translated into much longer operating life and correspondingly higher reliability for all elements of the spacecraft, particularly for the electronics.

But Pickering's "team" thrived on such challenges. As Pickering had so often asserted in public, it was challenges such as these that attracted the best of the country's engineering talent. Even within the Laboratory, the intellectual and technological challenge of the Mariner project attracted more of the better engineers at the expense of the less glamorous Ranger project.

In 1963 and 1964 new designs, some based on what had worked successfully before, others based on new and innovative ideas and techniques, were translated into hardware and software as the new breed of Mars spacecraft gradually took shape in the labs and test facilities at Pasadena. Quality assurance of the highest order, formalized failure-reporting and closeout, life and vibration testing, verifiable performance margins, and end-to-end system testing became enduring features of engineering life on the project. Over it all, Jack James' pervasive management style kept the project moving steadily forward, without the need for Pickering's intervention.

Two weeks after Pickering returned from South Africa, Mariner 3 left the launch pad at Cape Kennedy bound, regrettably, not for Mars, but for ignominious extinction. Within minutes after lift-off, the tubular shroud designed to protect the spacecraft during its passage through Earth's atmosphere, failed to disengage from the spacecraft as programmed and, in so doing, initiated a sequence of events that ended hours later with a dead spacecraft and a total failure of the mission. Pickering was appalled to say the least, although it was by no means clear at the time what had caused the shroud failure. Within days, James and his engineers traced the failure to the unvented honeycomb sandwich structure of which the shroud was constructed. Apparently, it had not been adequately tested under the conditions of heat and vacuum that it experienced as the launch vehicle passed through the upper atmosphere.

With the "launch window" for Mars due to close in less than 30 days, James made a decision to substitute an all-metal shroud for the original fiber-glass honeycomb version. Whether a new shroud could be built in the time remaining before the 1964 "launch window" closed, stood as an open question, but did not deter the team from trying. It did not remain an unresolved question for long.

Just 17 days after the first shroud failed, a replacement all-metal shroud arrived at Cape Canaveral for attachment to the Mariner 4 launch vehicle. While a second shroud went for accelerated testing, the countdown for Mariner 4 got under way at the Cape. The brilliant recovery from this crisis demonstrated the Laboratory's expertise at its best. The tests revealed no further flaws, and Mariner 4 eventually launched without problem on 28 November. All being well, Mariner 4 would arrive at Mars in mid-July the following year and return mankind's first close-up pictures of that most intriguing of planets. What would it see—Martians, canals, craters, or oceans? No one knew, but many were anxious to find out and theories abounded.

While JPL's latest dramatic production began to unfold in deep space, Pickering closed out the year with submission of a lengthy paper on "The Ranger Program" for publication by the AIAA, and an address on "The Surveyor Project" to the Management Club at the Hughes Aircraft Company in Los Angeles.

Although the public recognition that grew around William Pickering began, inevitably, to influence the lifestyle of both him and his wife, he never allowed it to dominate their private lives.

Muriel remained the warm, gentle, supportive personality that she had always been:

> . . . she took her public role very seriously. She certainly had parameters for herself as a Caltech wife . . . and later . . . as the wife of a famous person leading JPL. All the major events were a matter of careful planning, intense review, and analysis. Her approach was far more serious (than that of her husband). With Dad, the review of the events had always been from the point of view that the most amazing things happen and it's all incredible and often quite amusing. With Mom, it was conduct and correctness and protocol.

Perhaps it reflected a difference in their level of confidence to deal with the situations with which they became publicly engaged. His daughter agreed:

> Certainly, my Dad was confident in his knowledge and didn't need to operate with any pretensions. He loved what he did, and he was good at it, and because he knew that others contributed to

all the successes he acted with humility. He was comfortable with who he was and what he was doing.[34]

The controversy and criticism that surrounded William Pickering in 1963 and 1964 did nothing to dampen his enthusiasm for the space program and its overall objective to demonstrate the United States' preeminence in space technology. Nor was he deterred by the string of mission failures that were attributed to his leadership of the Laboratory in those years. Disappointed, yes, but not deterred. It was all part of the unavoidable learning process; "the steep part of the curve—just a part of the big picture," he would have said with a throaty chuckle and a twinkle in his eye. Moreover, he was firmly convinced that substantial progress had been made. His public lectures, especially those of a technical nature, were testament to his belief. Although there had been some successes with Mariner 2 and Ranger 7, the best was yet to come, and even if his detractors at NASA could not yet see it, Pickering sensed that redemption was just around the corner.

Endnotes

1 Folder 83 in the William H. Pickering Speech Collection, Pasadena, California: Archives and Records Center, Jet Propulsion Laboratory, JPL-181, 2004.

2 Folder 84 in the William H. Pickering Speech Collection, Pasadena, California: Archives and Records Center, Jet Propulsion Laboratory, JPL-181, 2004.

3 Folders 212-247 in the William H. Pickering Office File Collection, Pasadena, California: Archives and Records Center, Jet Propulsion Laboratory, JPL-186, 2004.

4 Folders 85-96 in the William H. Pickering Speech Collection, Pasadena, California: Archives and Records Center, Jet Propulsion Laboratory, JPL-181, 2004.

5 Mudgway, Douglas J., Oral History Interview with William H. Pickering, Part 6B, Pasadena, California, July 2003.

6 Folder 95 in the William H. Pickering Speech Collection, Pasadena, California: Archives and Records Center, Jet Propulsion Laboratory, JPL-181, 2004.

7 The records note that the only copy of this speech is a transcript from an AIAA tape recording that ran out of tape before the speech concluded. It is assumed therefore, that this speech was delivered without a fully prepared text.

8 Mudgway, Douglas J., Oral History Interview with William H. Pickering, Part 6B, Pasadena, California, July 2003.

9 Koppes, Clayton R., *JPL and the American Space Program: A History of the Jet Propulsion Laboratory* (New Haven: Yale University Press, 1982), p. 136.

10 Newell, Homer E., *Beyond the Atmosphere: Early Years of Space Science* (Washington, DC: NASA SP-4211, 1980), p. 265.

11 Mudgway, Douglas J., Oral History Interview with William H. Pickering, Part 6A, Pasadena, California, July 2003.

12 Koppes, Clayton R., *JPL and the American Space Program: A History of the Jet Propulsion Laboratory* (New Haven: Yale University Press, 1982), p. 148.

13 Ibid, p. 149.

14 Newell, Homer E., *Beyond the Atmosphere: Early Years of Space Science* (Washington, DC: NASA SP-4211, 1980), p. 267.

15 Koppes, Clayton R., *JPL and the American Space Program: A History of the Jet Propulsion Laboratory* (New Haven: Yale University Press, 1982), p. 151.

16 Mudgway, Douglas J., Oral History Interview with William H. Pickering, Part 6B, Pasadena, California, July 2003.

17 Ibid.

18 Hall, R. Cargill, Lunar Impact: A History of Project Ranger (Washington, DC: NASA SP-4210, 1977), pp. 223–252.

19 Ibid.

20 Kluger, Jeffrey, *Journey Beyond Selene; Remarkable Expeditions Past Our Moon and to the Ends of the Solar System* (New York: Simon and Schuster, 1999).

21 Hall, R. Cargill, *Lunar Impact: A History of Project Ranger* (Washington, DC: NASA SP-4210, 1977), pp. 223–252.

22 Mudgway, Douglas J., Oral History Interview with William H. Pickering, Part 6B, Pasadena, California, July 2003.

23 Hall, R. Cargill, *Lunar Impact: A History of Project Ranger* (Washington, DC: NASA SP-4210, 1977), p. 252.

24 Ibid, p. 256.

25 Koppes, Clayton R., *JPL and the American Space Program: A History of the Jet Propulsion Laboratory* (New Haven: Yale University Press, 1982), p. 163.

26 Hall, R. Cargill, *Lunar Impact: A History of Project Ranger* (Washington, DC: NASA SP-4210, 1977), p. 273.

27 Ibid, p. 278.

28 See Mudgway, Douglas J., Oral History Interview with William H. Pickering, Part 6D. Pasadena, California, July 2003.

29 On close inspection, it is apparent that the essentially "technical" speeches were written by Pickering himself; the more general, politically or philosophically oriented speeches were probably crafted by Wheelock, although there is no evidence to confirm that.

30 Folder 111 in the William H. Pickering Speech Collection. Pasadena, California: Archives and Records Center, Jet Propulsion Laboratory, JPL-181, 2004.

31 Folder 114 in the William H. Pickering Speech Collection. Pasadena, California: Archives and Records Center, Jet Propulsion Laboratory, JPL-181, 2004.

32 The tracking station near Johannesburg was one of three that comprise the Deep Space Network. The other two are in Woomera, Australia, and Goldstone, California.

33 Originally called Cape Canaveral to identify its physical location, the launch site was renamed Cape Kennedy in 1964 to honor the late President.

34 Beth Pickering Mezitt, Personal Recollections of my father, William H. Pickering. Private correspondence with the author, May 2003.

Chapter 7

Point of Inflection

Moon (1965–1968)

By the middle of the third quarter, William Pickering had lost interest in the game. The first half of the 1965 Rose Bowl college football championship match had been very one sided—the second half was turning into a rout. He recalled the events of a couple of years earlier when, as Grand Marshal for the 1963 Tournament of Roses, he and his family had been the guests of honor at the game. Now, he simply enjoyed privileged seating for himself and his family and was happy to be able to share this part of New Year's Day with them. As the game wound down to its inevitable conclusion, 34 to 7 in favor of Michigan, Pickering's thoughts drifted to other matters, foremost among them, the Ranger situation.

Of late there had been a steadily rising crescendo of criticism of the splendid images from Ranger 7. Questions had been raised about the real scientific value of the Ranger 7 images, and it had been implied that some of the claims of "importance for Apollo" went beyond what the actual imaging data showed. The Ranger imaging team had vigorously defended its interpretation, but after all the effort and expense, Ranger's worth to science, and to Apollo, remained unclear. As Cargill Hall noted, "To be sure, the solitary experiment remaining aboard the Ranger spacecraft, the visual imaging television system, had fulfilled the lunar mission objective specified by NASA. But the first close-up pictures had seemed to generate as much heat as they did light."[1]

And now, as Pickering slowly escorted his family through impatient crowds back to the Rose Bowl parking lot, Ranger 8 stood, in all its pristine beauty, not two miles away in the "clean room" at JPL, awaiting the decision of the "buy-off" committee to ship it to Cape Kennedy—the next step in another appointment with the Moon. Would it pass the intense scrutiny of the NASA-appointed committee and, if it did, would it actually make it to the lunar surface? Or was the Ranger 7 success merely a fluke, the result of a 1 in 7 chance of success? No one would know for sure until mid-February when, if all went well, Ranger 8 would transmit its final images to Earth in the last few minutes before it impacted the lunar surface.

Meanwhile, William Pickering faced a crowded schedule for the new year that began with his induction to the newly-founded National Academy of Engineering which had been established the previous year as a fourth element of the National Academies. As a founding member, he joined the company of the most prominent engineering professionals of the time: Terman, Ramo, Everitt, Millikan (Clark), Dryden, and Bode, to name a few. Together with his membership of the National Academy of Science he was becoming a significance presence in the prestigious world of American high science and engineering.

Right after a quick trip to Washington to consummate the formalities, Pickering embarked on a lengthy lecture tour of New Zealand. Sponsored by the University of Auckland, the 11th New Zealand Science Congress took place in February 1965, high summer for that part of the world. Distinguished scientists, principally from British affiliated countries took part. The Congress began with a public lecture in the Auckland Town Hall featuring "New Zealand's own space scientist," Dr. William Hayward Pickering speaking on "The Exploration of the Moon."[2] For the New Zealand public-at-large the topic was irresistible—it was a "standing room only" event.

Ever since the previous October when he had accepted the invitation to attend the conference, he had been deluged with invitations to address numerous local scientific and engineering organizations throughout New Zealand—all of which he, naturally, accepted.

The local media reported his every word with front-page headlines and pictures and comments. "N.Z. Born Scientist Has Big Role in U.S. Space Flights" exclaimed the *N.Z. Herald*; "Rocketing Posers His $200m Task," screamed the *Auckland Star*; "World Gets Benefit of Space Race," shouted the *Christchurch Star*; "Flying Saucers Reports Are Dismissed," reported the *Dunedin Evening Star*; and, in Wellington, the *Evening Post* announced loftily "U.S. Space Expert Visits Old School-Pupils thrilled to hear talk by famous Old Boy." Interviews with Muriel were immensely popular with the media and served to create a softer image of the world famous scientist who talked "their" language and really was "their hero." The Pickerings' visit gave rise to a remarkable outpouring of public interest and national pride in what were perceived as his personal achievements in a world beyond most people's furthest imagination. The Pickerings even found time for short family reunions with some of the Hayward family in Christchurch and members of the Pickering family in the Auckland area. The Institution of Professional Engineers N.Z. (IPENZ) took advantage of his visit to elect him to the rank of Honorary Fellow and the Governor-General invited Pickering and his wife to lunch.

Following his every move, the press suddenly reported that Dr. Pickering would be cutting his stay in New Zealand short. "My colleagues back in

Pasadena want me back earlier," he said. A few days later, in Christchurch, Pickering brought a gasp of amazement from his audience when he apologized, "I am sorry I am unable to stay longer in New Zealand, but I learned this morning that Ranger 8 is on its way to the Moon and I want to be back in the United States before it gets there." Who else in the world could say that? wondered those present. Pickering's words made them feel they were part of the drama unfolding on the other side of the world. To his adoring countrymen it was all part of Pickering's charisma—his enduring public appeal.

Just over 24 hours later, Ranger 8 unerringly executed its final sequences and, right on target, delivered thousands more dazzling close-up images of the lunar surface to its anxious controllers at the Goldstone tracking station during the final 24 minutes of its 248,000-mile journey. And by then of course, Pickering was back in Pasadena to preside over the momentous event.

Now it was the turn of the American press to laud the achievements of William Pickering. The *Pasadena Star-News* headlined "6000-Photo Finish Marks Ranger Trip-Jet Lab Jubilant on Shot." Speaking to a gathering of press, radio, and television reporters in the Von Kármán auditorium Pickering said, "It was a great satisfaction to see the project go so smoothly."[3] Scientific opinion based on quick reviews of the images suggested that the surface would be safe for an Apollo landing. While recognizing that a Russian spacecraft had already photographed the far side of the moon from a distance of 40,000 miles, Pickering believed that the Ranger images were far superior to those taken by the Russian spacecraft from a distance of 40,000 miles above the back surface of the Moon.[4]

The hint of justifiable pride in Pickering's remark was inescapable; Ranger 7 had not been a fluke, or just a matter of chance, after all. Perhaps, Pickering might have thought, the learning curve is flattening out—at last JPL is approaching the "point of inflection," as it were.

Whatever his innermost thoughts may have been, there was little time for pursuing them further for less than two weeks later he was due in Paris, France, to receive "Le Prix Galabert d'astronautique." Awarded annually by the Fédération Internationale d'Astronautique (IAF), an exclusive institution of which he was also president, the prize was also accompanied by a cash award of 7,000 Francs. Russian cosmonaut Valentina Terechkova received a similar award. French newspapers described him variously as " . . . père des engins spatiaux Mariner et Ranger,"[5] and "Le père des photos lunaires U.S."[6] and, naturally, reported the comments of Le Pr. Pickering in French, " . . . si la vie existe dans le système solaire c'est sur la planète Mars."[7] For his formal address to the IAF Bureau on 4 March he presented the "Exploration of the Moon" speech that he had given in Auckland a few weeks earlier.[8]

No doubt the French audience would have received his presentation with enthusiasm equaling that of its reception in New Zealand—it was a subject of universal interest at the time.

Early the following week, Pickering was in London, England, as the guest of the prestigious British Interplanetary Society. Founded in 1933, the Society was devoted to promoting the exploration of space and astronautics. In the course of a short visit, Pickering again presented "Exploration of the Moon" to an equally receptive audience of scientists and engineers. The following day he was back at JPL. He had been away from Pasadena for just six days.

Among the mail waiting his return he found an invitation from the National Space Club requesting the pleasure of his company at the Robert H. Goddard Memorial Dinner in Washington on 19 March. The featured speaker would be Vice President Hubert Humphrey. The occasion would celebrate the award of the annual Robert H. Goddard Memorial Trophy for 1965 to William H. Pickering " . . . for his leadership of the Ranger 7 Team in obtaining the first close-up pictures of the Moon." The trophy, a sculptured bust of Robert H. Goddard who was generally recognized as the father of modern rocketry, was presented by Hugh Dryden, Deputy Administrator of NASA.

Meanwhile, Ranger 9, which had completed its assembly and checkout tests at JPL, was now in the final stages of its prelaunch checkout at Cape Canaveral.

Morale and enthusiasm at the Laboratory had changed enormously in the past few weeks. Confidence in Ranger 9 and its design was at an all time high and expectations for yet another successful mission to the Moon were unbounded. Pickering later recalled:

> Ranger 7 and 8 had showed the extent of worldwide interest in what we were doing, particularly the photography, which was a 'natural' for the media. The fact that we told them for Ranger 9 we would give it [the television images] to them in real time was a demonstration of [our confidence] that it was going to work, and they were going to get something interesting. They were absolutely fascinated, no doubt about it.[9]

And he was right.

With the addition of appropriate scan converters to the existing data-processing equipment at JPL, it became possible to give the media a real-time television "feed" suitable for live broadcast on commercial television channels across North America. On 24 March, with Ray Heacock describing the lunar features as they appeared in rapid succession on the monitors in the Von Kármán auditorium, an awestruck audience of JPL personnel and media representatives and an unseen public television audience of many millions enjoyed the spectacle of "live television from the Moon" during the final minutes of the Ranger 9 mission to the Moon. It was a spectacular end to the first era of lunar exploration.

At the news conference that followed, Pickering observed:

> The project we reflect on today has been a long and difficult road since 1959. We had our problems in the early days . . . [but]

the achievements of the last three flights have shown that Ranger could carry out these deep space missions under remote command, that Ranger has indeed demonstrated the soundness of the basic system design and the close-up photographs . . . have opened a new field of the exploration of the Moon.

Historian Cargill Hall noted that, "Most could agree with those observations including Earl Hilburn, whose congratulatory telegram was to be found among others fast arriving at the Laboratory."[10]

From New York to Los Angeles the press acclaimed the world's first demonstration of "live television from the Moon" as Ranger's most impressive achievement. " . . . a front row seat on science," "Astronomy for the masses," and "High and historic drama," were typical comments to be found in the newspapers across the land. President Johnson, too, had been impressed by the television spectacular of Ranger 9 and, inspired by the additional success of the manned *Gemini* spacecraft a few days earlier, issued a public statement in which he declared, "Ranger 9 showed the world further evidence of the dramatic accomplishments of the United States space team . . . ."[11] Along with *Gemini* astronauts Grissom and Young, Schurmeier was summoned to the White House a few days later to be honored for his contribution to the exploration of space with the award of NASA's Exceptional Scientific Achievement Medal.

Despite the unequivocal successes of Rangers 7, 8, and 9, funding for the six follow-on missions that were intended to carry the real science payloads, was diverted to Apollo. Pickering was somewhat philosophical about NASA's decision to discontinue the Ranger project. He remembered:

> That was not an unreasonable thing to do. . . . If I were running the NASA program . . . and, I could put science on Apollo by canceling other programs, I would do it. Also I think from the point of view of public value we had just about milked Ranger for all it was worth with Rangers 7, 8, 9.[12]

With the cancellation of the remaining six Ranger missions, JPL's lunar program turned its attention to the Surveyor, lunar-lander project. In addition to a huge technological challenge, Surveyor brought a new management problem to the Laboratory. It was to be built under JPL contract supervision by a highly-regarded, aerospace contractor Hughes Aircraft Corporation. But the contract had not gone well, and by early-1965 when the spotlight moved away from Ranger and on to Surveyor, the Hughes contract had become a source of great concern to William Pickering.

Nevertheless, with the launch of the first Surveyor still one year away, and the work at Hughes gathering momentum, Pickering focused his attention on events in the more immediate future—events for which he held no ambivalent

JPL Director William Pickering indicates the impact of Ranger 7 on the lunar surface (Photo: NASA/JPL-Caltech Archives, Photo number P3412B).

feelings whatsoever—foremost among them, the progress of Mariner 4 arrival at the planet Mars, was then just three months away.

Important as it was, the first Mariner encounter of Mars was temporally preempted for Pickering's attention by the imminent marriage of his daughter Elizabeth Anne to Robert Wayne Mezitt of Massachusetts. The wedding itself was carried out "flawlessly" as Pickering would have said, as if describing a planetary encounter, and both he and Muriel were delighted with their daughter's choice of husband.

After the wedding William Pickering hastened back to California for another appointment with history. On 14 July, Mariner 4 arrived at Mars and completed its mission "flawlessly" by transmitting 21 television images of the Martian surface back to Earth as it passed by the planet at a distance of 10,000 miles above the surface. The images, covering about 1 percent of the planet's surface, were transmitted from Mars to the tracking stations of JPL's Deep Space Network at a rate of approximately eight bits per second, a major feat in deep space communications for its time.[13] As the numbers, representing light and dark areas of the Mars surface, arrived at JPL by teletype from the distant tracking stations, jubilant scientists pasted the tapes in strips on a large card and colored the like areas to produce the first digital image of Mars's surface features. Later, JPL's science

Pickering family group at the wedding of Elizabeth Anne Pickering to Robert Wayne Mezitt, Ithaca, New York, June 1965 (Photo: Courtesy of Pickering Family Trust).

team presented the crude image to the Director with great pride. Eventually the crude image became a much-valued artifact of a historical "first" in deep space and, incidentally, an icon of Pickering's long-held ambition to be "first in deep space." It took a while longer to produce the detailed images that the scientists required for their analysis of the Mars surface features.

Again, the media hailed William Pickering as the man behind yet another, amazing achievement in space.

The *New York Times* carried a piece by Walter Sullivan on the front page titled "Mariner 4 Makes Flight Past Mars." While noting that the spacecraft had already determined the absence of a significant magnetic field on Mars, the article pointed out that it would take over 8 hours for the spacecraft to play back a single image of the planet surface features at just over eight bits per second.

At that rate, it would take about 10 days for all the images to trickle back to Earth across the vast emptiness of space. At the press conference Dr. Pickering had warned the media not to expect too much from the Mariner television system. However, he said these would be "the first ever obtained in the vicinity of another planet," and, " . . . should be better than any ever obtained from earth if the system operates properly." Among many other things, the pictures

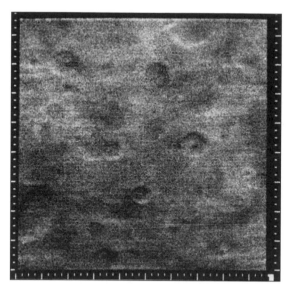

Mariner 4 picture of Mars surface (Photo: Available online at *http://nssdc.gsfc.nasa. gov/imgcat/midres/m04_09d.gif*).

should lay to rest the age-old controversy over the existence of Martian canals that underlay early beliefs of an advanced Martian civilization.[14]

In addition to Sullivan's leading article on the Mariner 4 encounter with Mars, The *New York Times* also carried a handsome personal tribute to Pickering titled "Expert on Spacecraft; William Hayward Pickering." The article described JPL and Pickering's early career, of course, but it also captured the quintessence of Pickering's advocacy for the national space program, and its importance to the nation's international stature. It suggested that his "greatest contribution" may have been his positive efforts to influence government and public attitudes toward support for the space program, and lauded his determination to rally public confidence in the nation's power to recover from the shock of Soviets' dominance in space engendered by the Sputnik affair and subsequent Moon shots. It commended Pickering's belief in "strong leadership and good engineering management," rather than spectacular space events of shallow scientific significance, and his unshakeable opinion "that the most important aspect of the space race was international prestige." The writer observed that Pickering called for an end to military competition for space projects, and urged a truly unified national space program " . . . with both military and civilian research coordinated under NASA."

Finally, the article recognized the impressive standing that Dr. Pickering bore as an honored member of many national and international scientific organizations, as evidenced by the great esteem with which he was regarded by his colleagues. But, said one, "None of these honors can add luster to the stature he has acquired in 'nearly accomplishing the impossible'."[15]

All of these opinions and comments reflected the impact that Pickering had generated with his plethora of public speaking, published articles, and widespread influence in professional societies of all kinds, as we have seen.

A week later, with the first several Mars pictures in hand, *Time* magazine recognized William Pickering for the second time with a front page cover titled "Mariner's William Pickering."

The *Time* article titled "Portrait of a Planet," told of first impressions of the images. Although the pictures were not all that great—"grainy and ill-defined" and it would require much more analysis to interpret their real import—they conveyed a most important message " . . . from 135 million miles in space . . . Mariner 4 had sent home the first close-up portrait man has ever made of far-off planet Mars." The other photos were similar and showed extensive desert-like areas, with a few indistinct surface features that suggested shallow Moon-like craters and elongated depressions. In addition to the images, the scientific return from Mariner included important data on the fields and particles of the deep space. All of this data had been returned to Earth over the longest deep space communications link ever achieved.[16]

The *Time* article, attributed the success of the Mariner 4 mission to Mars to " . . . one of the most skilful and resourceful teams ever gathered together in pursuit of scientific knowledge." Alluding to Pickering's difficult times with his NASA critics, *Time* said, "These men were part of the arrogant, egotistic, brilliant, experienced, and single-minded team on which William Pickering staked his career, and his reputation, when he defended his management style to those who would have it otherwise, but who were now the beneficiaries of its success." Acknowledging Bill Pickering's leadership, *Time* concluded "these men have fashioned the most ambitious and successful space venture yet." It was indeed another handsome tribute that did much to vindicate Pickering's faith in himself, his style of leadership and the capabilities of his "team" at JPL.

Although these pictures would add little to the question of life on Mars, Pickering was optimistic about the future, "I've always felt we'll find some form of life on Mars, and I look forward to the day when we are landing capsules there and searching for life," he said.[17]

Neither William Pickering, nor anyone else for that matter, could have known how prescient that observation really was. Just ten years later, NASA and the JPL team would be doing exactly that—searching for life on Mars with a huge spacecraft called Viking.

The Mariner 4 mission left a lasting impression with William Pickering. "The images showing that Mars is more like the Moon and not like the Earth was an important scientific achievement," he recalled, "because there were a lot of people who were inclined to believe . . . that there was some kind of civilization on Mars. But these [pictures] proved definitely that there was a difference between Mars and Earth."

Perhaps more than anything else, the Mariner 4 mission firmly established United States' lead in the race with the Soviets, a sentiment that would be echoed from the White House a short time later when the President again summoned Pickering, together with Oran Nicks, Director of NASA's Lunar and Planetary Programs Office, to receive NASA's Distinguished Service Medal.

For Pickering, Mariner 4 represented a transition from the developmental to the operational phase in the evolution of planetary spacecraft.[18] As Pickering put it, "We now knew how to do it." JPL had reached the "point of inflection" on Pickering's learning curve.

In the last four months of 1965, Pickering made a dozen major speeches and presentations. Muriel accompanied him on some of the longer trips. Together they visited Athens, Greece, to open the XVIth Congress of the International Astronautical Federation (AIF); Kauai, Hawaii, for the Governor's Conference on Oceanography; and the International Space Electronics Symposium in Miami, Florida. He delivered a paper on the "Mariner Flight to Mars" for the journal *Astronautics and Aeronautics* in October, and another on "Some New Methods for Planetary Exploration" for the National Academy of Sciences fall meeting in Seattle, Washington.[19] In the latter speech he spoke of a new application for radio astronomy where the radio "beam" from a very powerful transmitter at the Goldstone tracking station had been pointed at distant planet Venus. Careful interpretation of the characteristics of the radio signals reflected from the planet had revealed much hitherto unknown scientific knowledge about the dynamics of the planet and its surface.

Turning to robotic spacecraft, Pickering said that within the past five years it had become possible to conduct scientific experiments and make scientific observations from the vicinity of the planets themselves by making use of robot

spacecraft to carry the instruments to the desired locations. Scientists from both the U.S. and the Soviet Union had demonstrated these techniques with varying degrees of success, but both countries would undoubtedly continue these efforts. Pickering illustrated all of his salient points with examples from Mariner 4, and explained how the telemetry data was received

Wearing the NASA Distinguished Service Medal, William Pickering explains the pictures of Mars's surface taken by Mariner 4 to President Johnson: August 1965 (Photo: NASA/JPL-Caltech Archives, Photo number P5109A).

at the tracking stations in California, Australia, and South Africa, and passed to JPL for processing and scientific analysis. Mariner 4 was the most sophisticated planetary spacecraft yet launched by the U.S., but the next generation of planetary spacecraft would be much larger in both weight and capability. "Present plans call for orbiting and landing versions of Voyager, starting in 1971, for Mars exploration," he said.

The speech was a masterful summary of the state-of-the-art in planetary exploration, and presaged future developments in deep space exploration that, although threatened by the voracious fiscal appetite of the mighty Apollo program, nevertheless survived and eventually came to pass in ways that he could not then have imagined. William Pickering spoke from a position of increasing strength, obviously secure in the knowledge that the space program as he saw it was making good progress in the right direction.

At the end of that year he could regard with some pride, and not a little wonder for he was a modest man, the collection of prestigious medals and prizes, some accompanied by significant monetary awards, which had been bestowed upon him by prominent institutions in the field of aerospace related technology. From France came the Prix Galabert and from the U.S. Space Club came the prized Goddard Memorial. U.S. Army Ordnance presented the Crozier Gold Medal and the U.S. Army elected him as their Citizen of the Year for 1965. The American Society of Mechanical Engineers awarded him its "Spirit of St. Louis Medal" for 1965, while the Scientific Research Society of America gave him its Proctor Prize for that year. And it was in that remarkable year also, that President Johnson presented him with NASA's Distinguished Service Medal.[20] Almost all of these awards were based on his outstanding contribution in one way or another to the field of "space exploration." Although he had not actively sought recognition, he accepted it all graciously and, in most cases, responded with a personal appearance and an appropriate speech of acceptance.

Interpreted loosely in Spanish as "a shady glen," La Cañada, to use its original Spanish spelling, was one of several small communities that originated as orange orchards along the foothills of the western San Gabriel Valley at the turn of the 20th century. By the 1960s the orange groves had gone and La Cañada had become an equestrian-oriented community of larger, upper-class, country homes whose wealthy owners regarded the presence of the Jet Propulsion Laboratory in the nearby arroyo as an intrusion into their exclusive community.

By that time, several large, sparkling new buildings had appeared within JPL's campus. The successes of Ranger and Mariner enhanced the reputation of JPL as an attractive place to work, and the population of the Laboratory increased rapidly. Scientists, engineers, technicians, secretaries, administrators, and support staff quickly filled the vacancies at JPL as they became available.

Newcomers moved into the area and La Cañada became a highly desirable place to live and to raise a family.

On the broad flanks of the hillside overlooking JPL to the north, the fashionable community of Flintridge developed almost as a mirror image of the big homes that comprised the original La Cañada enclave across the valley. This was the location that William Pickering chose for his new home. With a breathtaking view on all three sides, a swimming pool, and a direct view down to the rapidly expanding JPL campus, it was a natural choice for the public figure that William Pickering had now become. Pickering's helicopter, exiting the JPL landing pad en route for the airport, passed his home at eye level, affording, on occasion, an opportunity for a quick wave to Muriel.

Both he and Muriel quickly settled in to the social life of the new community. Muriel established close associations with many of the civic groups in La Cañada. For his part, William took an active interest in the educational and local government groups that had begun to shape the character of the rapidly evolving city of La Cañada-Flintridge.

The Surveyor lunar landers were strange, three-legged machines that, rather than crashing violently as the Rangers had done, would descend gently to the surface. They would deliver video and science measurements in real-time from the Moon to engineers and scientists at JPL. Later versions were to be equipped with a miniature trenching tool that scientists would manipulate remotely from JPL to reveal the nature of the lunar material to a depth of a few inches below the surface, turn over rocks to reveal what lay beneath and generally test the mechanical properties of the lunar soil. Other instruments would test the chemical composition of the soil. All of these data, in addition to close-up photographs of the near and distant lunar landscapes, would stream continuously to Earth during the lunar daytime when solar power would be available. The findings of these complex spacecraft would surely provide the Apollo mission designers with the critical details of the strength of the lunar soil on which the safety of the manned landings depended. As a minimum, the Surveyor data would confirm their assumptions, and extend the predictions that had been based on the data from the three Rangers.

But there were problems—not so much with the Surveyor spacecraft, but with the Surveyor contract.

Somewhat paralleling the problems that beset JPL in its early Ranger days, the problems with the Hughes contract stemmed from inadequate planning, the nature of a cost-plus-fixed-fee contract, inadequacy of the company's management, and technical infrastructure to handle a project of this complexity, and lack of sufficient supervision by NASA and JPL.[21] The previous year Pickering had been forced by a fully empowered NASA review into designating Surveyor as the "top priority activity of JPL" and instituting a massive "rescue operation" that involved assigning as many as five hundred of the Laboratory's most experienced personnel to work full time on Surveyor; the author was one of them. Under these extreme measures, the technical and engineering status of the contract slowly improved, but contractual problems persisted.

Into this vexatious situation stepped General Luedecke, Pickering's Caltech-appointed deputy. Almost immediately, Luedecke invoked his authority, business acumen, and empathy with NASA Headquarters to discipline the Surveyor cost accounting processes.[22]

If Pickering felt ambivalent about Surveyor, he left no record of it, and there is no suggestion that he gave it less than his full support. To all outward appearances, JPL's problems were now behind it; having, at least, equalized the race for primacy in space, the Laboratory could look forward to even greater triumphs in space in the years ahead. Such was the image that Pickering's frequent and wide-ranging public appearances created in the minds of those who heard him and those who read the widely published reports of JPL's spectacular successes.

These perceptions were further enhanced when, in June 1966, the first Surveyor spacecraft landed gently on the surface of the Moon, activated its cameras and, without fuss or bother, began transmitting the first pictures of the lunar soil beneath its feet. It was accomplished so easily that it almost seemed a matter of routine, rather than being a technological feat of the first order. Surveyor had landed upright, its three landing pads penetrating the lunar soil to a depth of an inch or two with all of its systems in perfect working order.

For the next month Surveyor 1 responded faultlessly to over 100,000 commands from its Earth-bound controllers. It returned more than 11,000 images of the lunar landscape including rocks near and far and material around them, and, for the first time, enabled Apollo mission designers to see exactly what their manned landing craft and its occupants would face when they eventually touched down on the lunar surface. And once again it had become a media "feeding frenzy."

Pickering remembered the occasion well:

> There was a lot of media excitement about the first one. They had gotten used to real-time coverage on Ranger and they wanted to do the landing in real time, but I did not want them to do that because we were pretty unsure about the success of this thing. But they did an end run around me and we got orders from Washington to, 'Give it [real-time television of the landing] to the media.' So that is what we did . . . and they spread it all around the world. It was just as well it was a success.

When asked about the Apollo reaction to the Surveyor success he added somewhat ruefully: "I felt the Apollo people should have been more interested in this than they were, but they said 'Good landing,' that's all."[23] Astronaut-scientist Harrison Schmitt thought differently, "Ranger and Surveyor were of tremendous importance," he replied when asked how valuable they were to the manned Apollo missions adding, "There was a more political question, I think, than geological question about the nature of the surface."[24]

A few weeks later, a *Life* magazine article waxed eloquently over the Surveyor pictures and what they represented, "The color photographs constitute a crowning achievement in Surveyor's success . . . it carried out its task so efficiently that even its most optimistic designers were dumbfounded."[25] Although the Soviet spacecraft Luna 9 had preempted Surveyor by landing successfully on the Moon a few months earlier, it had survived only three days and returned only nine lunar images, a fact that did nothing to minimize the elation over the Surveyor's success.

Inevitably, some of the credit for the successful mission was attributed to General Luedecke's influence on the contract relations with Hughes and JPL, and NASA Headquarters viewed his efforts with great satisfaction.[26] But rather than build upon this perception to improve his relationship with Luedecke, Pickering, rather inexplicably, elected to curtail his deputy's influence by sharply limiting his authority. Pickering issued new job descriptions that reduced the authority of his deputy director, while simultaneously increasing the authority of his several assistant directors who were responsible for technical and administrative work of the divisions.

As a consequence of this action, the balance of executive power at JPL effectively reverted back to where it had been prior to the time that NASA had forced Caltech to appoint a deputy director for JPL. Not surprisingly, Luedecke viewed this as an affront to his authority and appealed to NASA for help to resolve, what was for him, an impossible situation.

For the next year, a stream of accusations, denials, protests, and complaints surged back and forth between the four parties, NASA, Caltech, Luedecke, and Pickering. Eventually, Pickering took the position that Caltech would have to choose between him and Luedecke, and that while Luedecke remained in office, he could no longer continue as Director. To back up his argument, Pickering reminded DuBridge of his allegiance to JPL and concerns that JPL could well become "merely a job-shop for NASA."[27] There could be little doubt that Pickering was voicing his most deeply felt concerns and putting his career on the line to prove it.

Ultimately, Caltech's Board of Trustees upheld Pickering's position and General Luedecke quietly resigned and moved away to a position as president of a large university in Texas. The position, however, had not gone away. Shortly afterwards Caltech hired Rear Admiral John F. Clark to fill the position of deputy director at JPL. It was a happy choice. Admiral Clark quietly settled in to the deputy director position at JPL and was readily accepted by both the Director and the senior staff of the Laboratory. When Clark retired a few years later, Caltech continued the military tradition with the appointment of Air Force General Charles Terhune as deputy to Pickering. Pickering and Terhune eventually became great friends and created a productive management partnership that acted for the remainder of Pickering tenure, to the mutual benefit of NASA, Caltech, and JPL.

Despite the furor surrounding his interpersonal relationships at JPL in 1966, Pickering still managed to find time for international visits and speeches—Paris, France, "Why Go to the Moon?;" Madrid, Spain, "AIF conference;" London, England "Speech on the BBC2"—in addition to delivering ten or more speeches from one side of the U.S. to the other. Two honorary doctorates, one from Occidental College in California, the other from Clark University in Massachusetts, the Magellanic Premium from the American Philosophical Society, and a further foreign distinction of the meritorious Order of Merit of the Republic of Italy were indicative of the great esteem with which William H. Pickering was regarded in this country and abroad.

Excitement over the success of Surveyor 1 was short lived. Surveyor 2, launched about three months later, experienced a failure during its midcourse maneuver and crashed uselessly on the Moon a few hours later. It would be April the following year before the program resumed with the launch of Surveyor 3. In addition to their television cameras, Surveyors 3 and 7 were each equipped with a remotely controlled digging tool called a surface sampler. It could be used for soil mechanics experiments that involved trenching, scooping, and depositing lunar surface material in view of the camera. It could also be used to push around, or overturn, small rocks within range of its extendable arm. Magnets attached to the footpads of Surveyors 5, 6, and 7 tested the soil for its magnetic properties and an alpha-ray scattering experiment was used to make chemical analyses of the lunar soil.

Except for Surveyor 4, which disappeared without trace about two minutes before touchdown, all of the remaining Surveyors succeeded in reaching the surface of the Moon and carrying out their lunar experiments as planned. Close-up pictures of the lunar terrain, portions of the spacecraft, and crescent Earth viewed from the Moon became commonplace, as scientists struggled to interpret the avalanche of new data that flowed into their data banks.[28]

NASA's lunar lander program concluded in January 1968 with Surveyor 7, but by that time Pickering had set his sights on Mars and the planets that lay beyond. The Surveyor missions soon faded from public view and the world's attention became riveted on the unfolding drama of the Apollo manned lunar landings. Surveyor made a brief reappearance on the world stage in November 1969 when Apollo 12 astronaut Pete Conrad retrieved the camera from Surveyor 3 still standing peacefully where it landed gently in the Ocean of Storms more than two years earlier. Returned to Earth on Apollo 12, the camera from Surveyor 3 was closely examined, and to the amazement of its designers, determined to be in good condition. Eventually NASA presented it to William Pickering on his retirement, and it became a permanent exhibit at the Air and Space Museum in Washington, DC.

The appointment of a more compatible individual as deputy to William Pickering did much to ease the tensions in the Director's office at JPL, but did nothing to quench the NASA Administrator's drive to mold JPL and its recalcitrant Director to his vision of the ideal NASA/Caltech/JPL triad. No sooner had General Ludecke departed the Laboratory than Administrator Webb announced the formation of yet another NASA-sponsored committee to investigate JPL/Caltech management. The new committee was to be chaired by UCLA professor Chauncey Starr, a professional colleague who was well known to, and highly regarded by, Pickering. Rather as Pickering had anticipated, the Starr Committee's findings at the conclusion of its nine-month investigation of JPL and its interactions with NASA and Caltech were rather benign, and fell far short of the sweeping changes that Webb had looked for in initiating the investigation in the first place.[29]

The long shadow of Webb's influence swept over JPL for the last time a few months later during the 1968 contract negotiations. Once again NASA

Surveyor footpad and trench in lunar soil (Photo: National Space Science Data Center).

attempted to "diminish Pickering's authority and JPL's autonomy." As was to be expected, Caltech and JPL objected strongly and resisted most of NASA's demands. Before any of the new agreements could be put into effect, however, James Webb retired from NASA and was succeeded in October 1968 by the former Deputy Administrator Thomas Paine.[30]

Almost immediately the hitherto conflicted negotiations began to make progress. New guidelines designed to achieve "mutual objectives" were set, a fund to sponsor independent research at JPL was established, and Paine consented to a substantial increase in the Caltech "fee." And, what would later prove to be of great importance to Pickering, NASA permitted JPL to seek funding from "agencies other than NASA."[31] Both sides appeared satisfied with the ultimate and wide-ranging outcome of the negotiations.

Without the burdensome influences of Webb and Luedecke in his professional life, the future might well have looked rosy indeed to William Pickering —but that was not how it turned out.

Apollo 12 astronaut Pete Conrad retrieves camera from Surveyor 3 standing in the Moon's Ocean of Storms. Returned to Earth on Apollo 12 for analysis, the camera was eventually delivered to William Pickering (Photo: NASA/JPL-Caltech Archives, Photo number P10623B).

Planets (1967–1969)

As he soon discovered, the future held challenges that taxed Pickering's leadership abilities beyond anything he had encountered or could have expected. The technical and administrative challenges of the early 1960s were now superseded by an urgent need to diversify JPL's interests and attend to its very survival.[32] Changes were everywhere. The tremendous verve, excitement, motivation, and pride that drove the Laboratory staff to extraordinary accomplishments in the past began to dissipate as "deep space spectaculars" became commonplace and the exploits of the Apollo astronauts stole the public's interest. Other NASA centers, too, began to compete with JPL for NASA and public recognition, as their own deep space missions began to make headlines in the national media. The intrepid Pioneers from Ames Research Center, the highly successful Lunar Orbiters, and, later, the magnificent Vikings from Langley Research Center were prime examples.

Although he did not undertake so much international travel in 1967 and 1968 as he had done in earlier years, there was no let up in Pickering's frenetic drive to address institutions and organizations far and wide across the U.S. Fifteen public addresses in 1967, one of them in Belgrade, Yugoslavia, and thirty in 1968, including several in New Zealand and one in Genoa, Italy, surely represented a remarkable personal effort on the part of a man who, while directing a major technological institution engaged in the world's leading space initiative, was simultaneously engaged in career-threatening internal controversy with its principal sponsor. On top of all this, he continued to garner awards and honors from civic and professional organizations alike. The city of Pasadena, the Aeronautics and Astronautics Society, American Institute of Aeronautics and Astronautics, and Institution of Electrical and Electronics Engineers were among those that recognized his contribution to the new field of space exploration with their most prestigious awards.[33]

As they had done so often in the past, Pickering's speeches in 1967 and 1968 reflected his concerns for the present and future condition of the U.S. space program. In 1967 he frequently visited themes that reviewed the progress of the first decade since Sputnik: "Progress in Unmanned Space Exploration," "The Opening Decade of Space Exploration," and looked forward to the future with "The Next Steps in Space Exploration" and "The Next Ten Years in Space." By the following year, his evident concern for the shrinking space budget and the disproportionate allocation of available funds to the Apollo manned program at the expense of the unmanned exploration of the solar system became clearly apparent with "Why Explore the Solar System" and "Neo-Sputnik: The Age of Unreason in Space Exploration."

This latter speech, delivered to a group of businessmen at the Los Angeles Rotary Club in August 1968, revealed his concern for the present and his vision for the future of the space program. The broad sweep of the ideas he espoused in this speech stand in marked contrast to the petty aggravations that competed for his attention at JPL in those years.

"A decade earlier Sputnik shattered some long-held ideas among Americans, that the Soviets were inept scientifically and technologically . . . and the American public reacted with near panic," Pickering said. But after some hasty improvising and some early failures, the U.S. recovered to the point where, by mid-1968 it holds " . . . a comfortable lead in scientific innovation and technological productivity." He attributed much of the credit for this situation "to the expansion of government support of research and development." That had now changed. While the 1969 budget allocated $2 billion for Apollo, it only provided funding for, at best, " . . . a minimal planetary program."

"Can the nation maintain its world leadership in science and technology without vigorous support from the government and the public?" he asked. Citing recent polls that showed public interest in the space program to be at its lowest ebb since the beginning of the decade, he said that many respondents were willing to abandon planetary research to the Soviets. "Even landing astronauts on the Moon had lost much of its glamour." On the other hand, the Soviets, he claimed " . . . showed every sign of being eager to resume the lead in space exploration which . . . they consider a valuable weapon in the contest for men's minds," and could be expected to "launch to Mars and Venus at every opportunity."

"If we do not avoid . . . the type of irrationalism which ignores the hard lessons learned during the current turbulent decade," he warned, " . . . one day in the early 1970s a new Sputnik is likely to appear on our horizon."

"Must history repeat itself?" he concluded.[34]

Embedded within this compelling speech were clear references to powerful forces like the reduction in funding for the space program, dominance of the Apollo program for available funds, and diminishing public support for continued space exploration that would determine the character of the planetary program for the remaining years of Pickering's tenure with the Laboratory.

That it survived this hiatus in the nation's planetary space program is, in no small measure, due to Pickering's efforts to diversify the scope of the Laboratory's interests.

Toward the end of 1965, William Pickering had received an important message from NASA Headquarters. Although not unexpected, it was welcome nevertheless for it authorized him to initiate work on a new mission to Venus. This mission would take advantage of the 1967 Venus "inferior conjunction"[35] when Earth and Venus would be on the same side of the Sun and therefore the distance between Venus and Earth would be at a minimum. To save time and cost, the Mariner 5 spacecraft was to be constructed from surplus Mariner 4 flight-qualified components that had been carefully preserved at JPL. Only minimal changes to basic design of the spacecraft were permitted, and the science instruments were limited to a radiation detector and a magnetometer.

The most important science experiment, measurement of the density of the Venusian atmosphere, was to be determined by analysis of the changes to the spacecraft's own radio signals as the spacecraft moved behind the planet's disk as viewed by antennas on Earth, an astronomical condition known as occultation.

Despite the Laboratory's major emphasis on Surveyor, enough effort was found to build and launch the new planetary spacecraft on schedule. Mariner 5 reached Venus without incident, in October 1967, a little less than two years after Pickering received the "go ahead" from NASA. Although the encounter period of closest approach was only about one hour in duration, it was sufficient for the instruments to make their critical measurements; it was searching for the presence of a magnetic field and a belt of radiation, and measuring the height and temperature of the upper atmosphere. When the spacecraft disappeared briefly behind the planet as viewed from Earth, the radio "occultation" data, which held the key to the composition and density of the atmosphere, was recorded at the tracking stations for later analysis. It was a perfect mission in every sense.

The *New York Times* reported the event on an inner page with a comprehensive article by John Noble Wilford that quoted the Mariner project scientist in explaining the mission and the science results.[36] There was no mention of the JPL Director, as there had been on previous occasions. He was evidently standing back to give prominence to his project managers when the press was around. As his daughter commented, " . . . he was becoming self-conscious about getting all the praise when others did all the work."[37]

To some extent, the impact of Mariner 5's encounter with Venus was upstaged by that of a Soviet spacecraft, Venera 4, that had dropped a capsule on to the Venusian surface the previous day and, the Soviets claimed, had made measurements of atmospheric density and temperature from the actual surface for a short time before the capsule failed. This had made front page news in The *New York Times* the previous day but the Soviets' data had been called into question by the occultation data received from Mariner 5.[38]

This was the first demonstration of the powerful new technique of radio "occultation" that would become a key experiment on all planetary encounters in the future. Pickering was really impressed. Recalling the occasion with obvious pleasure. He said:

> We got into an argument with the Soviets over that [use of radio occultation data]. . . . Shortly after Mariner 5 went to Venus, there was an IAF meeting where the Soviets presented their claim that Venera 4 had landed but failed on impact. From this they calculated [a value for] the radius of the solid body of Venus—and Kliore [a prominent radioastronomy scientist at JPL] used his radio occultation data to show that they were wrong . . . the pressure of the atmosphere is what turned their

machine off. They did not believe us until we showed them Kliore's data . . . then they accepted it and beefed up their modules to withstand the pressure.[39]

Although the Soviet spacecraft had arrived at Venus one day ahead of the American spacecraft, Pickering derived lasting personal satisfaction from having proved some of their data in error.

"Spacecraft occultation," the powerful new type of radio science data that had been demonstrated so convincingly on the Mariner 5 Venus encounter, had been made possible by the recent addition of a gigantic new radio tracking antenna to JPL's Deep Space Network.[40] The new, 210-foot diameter antenna at Goldstone was the largest precision spacecraft-tracking antenna in the world, and embodied much new technology that would eventually enable future planetary spacecraft to reach the very edge of the solar system. There was only one for now, but two more would soon be built, one in Australia near Canberra and the other in Spain near Madrid.[41]

Except for the "occultation" experiment, Mariner 5 had not made full use of the additional capabilities afforded by the great new 210-foot diameter antenna at Goldstone, since its earlier spacecraft were designed to match the existing 85-foot diameter antennas of the Deep Space Network. But all of that was about to change as NASA again set its sights on Mars and ordered Pickering to begin work on the design and construction of two new spacecraft to unlock Mars's closely guarded secrets during the Mars 1969 opposition. These second-generation Mariner spacecraft would be designed to make full use of the giant new antenna to speed up the return of imaging science and engineering data from Mars to Earth and enable future spacecraft to penetrate even further, into the mysteries of deep space.

Compared to earlier versions, the second-generation Mariners were bigger, heavier, and carried a suite of greatly improved science instruments mounted on a scan platform. They also carried an improved television system, a central computer that could be reprogrammed in flight, an upgraded data storage system, a multiple channel telemetry system, and a capability to return data from Mars at 16,200 bits per second using the new Goldstone antenna. Mariner 6 and 7 would be the first of a new family of planetary spacecraft designed expressly for deep space exploration.

Both spacecraft experienced a number of problems, but eventually reached Mars within a few days of each other, toward the end of July 1969. In the media, the Mariner 6 and 7 encounters of Mars competed for public attention with the high level of residual interest from the Apollo 11 manned landing on the Moon that had taken place a week earlier. There could be little doubt that the United States' space program was in full swing.

In reporting the two events, The *New York Times* again quoted the project scientists rather than the Laboratory Director, and focused its attention on the

NASA/JPL's new 210-ft diameter antenna at Goldstone, California: April 1966 (Photo: NASA/JPL-Caltech Archives, Photo number 332-9278Bc).

technical and scientific aspects of the two missions. There was little or no reference to the space race or the U.S. preeminence that had characterized JPL's major space events a few years earlier.[42] The scientists found the presence of light and dark areas, the absence of "canals," the presence of polar snow caps, and gigantic surface features like Nix Olympia of particular interest, and characterized the surface as more Earth-like than Moon-like. The science team generally expressed uncertainty over possible existence of life. One writer noted the contribution of the Goldstone 210-foot (64-meter) diameter antenna, observing that it had made it possible to receive the Mariner data from Mars at 16,200 bits per second. Without it, the 85-foot diameter antennas would reduce the data rate to a mere 8 bits per second, meaning that each picture would take up to 6 hours rather than 5 minutes to reach Earth.[43] Pickering's telemetry system had come a long way from its earliest days and he fully appreciated the advanced deep space communications technology, mostly developed at JPL, that had made it possible.

If the Mariner 6 and 7 images of Mars raised more questions than they answered about the origin of the planet and the possibility of its harboring some sort of primitive life form, they also served to stimulate interest in a follow-on mission to Mars, this time using an orbiting rather than a fly-by spacecraft. And the next opportunity for a mission to Mars would come as soon as 1971.

Earth (1969–1976)

The absence of William Pickering as a prominent personality in the media depiction of the Mariner 5, 6, and 7 encounters of Venus and Mars was symptomatic of the changes then engulfing the Laboratory. The NASA budget, on which those events and the Laboratory's very future in space depended, had begun to diminish. By 1969, JPL's budget had fallen to little more than half of what it had been in the peak year of 1967, and NASA was calling for further substantial cutbacks in programs and staff. Ironically, just as the Laboratory appeared to be ascending the learning curve it was rapidly descending the funding curve. Both these facts were of profound concern to William Pickering. What was to be done?

Pickering's dilemma was, of course, but a reflection of NASA's overarching problem with diminishing Congressional support for space projects. NASA's approach took the form of building public awareness for "technology transfer," or "spin off," from space technology to the public sector. Pickering, however, took a slightly different route to preserve his beloved organization. Invoking the earlier agreement in the NASA contract that allowed JPL to seek funding from non-NASA sources, Pickering established a "civil systems" organization within JPL and set about looking for suitable projects and sponsors to support it. If JPL's expertise in the technologies of space was not fully extended in supporting the nation's space program then Pickering believed it could, or should, be brought to focus on solving some of Earth's most pressing social problems.

This topic had apparently been on his mind for some time. In December 1968 he had delivered a lengthy speech in Los Angeles titled "Science and the Urban Crisis-A Fragmented Dilemma."[44] Pickering attributed the failure of Washington, local governments, scientific, industrial, and financial communities to alleviate the urgent social problems now facing America's cities to "fragmentation of authority at the local level."

In the closing years of the 1960s, the nation was disturbed by unrest throughout every phase of its social structure—from rioting in the ghettos to divisiveness over the Vietnam war; from crises in education to concern over the economy. The prospects for the future, said Pickering, " . . . were bleak indeed, unless we get major social and cultural innovations during the next three decades." To do that we must marshal our very considerable resources in industry, finance education, and, most importantly, apply science and technology to our social problems. A technique that had proven most effective in developing our postwar missile and aerospace problems was known as "systems engineering." And, he believed, these techniques held promise for application in the field of social engineering also.

Pickering quoted Vice President Humphrey in saying that "Systems analysis can contribute importantly to community planning, police and firefighting services, educational systems, urban modernization, control of crime and delinquency transportation . . . effective use of natural resources and the elimination of water, air and soil pollution." But to be effective, these techniques must be initiated at the federal level. Otherwise, "fragmentation of authority at the local level" would eventually bring about their failure. "Since our urban crisis constitutes a challenge of overwhelming national urgency, science and technology must address themselves to this critical area if we are to make real progress in this century," he said.[45]

And so William Pickering set the stage for a new perspective on the proper function of his Laboratory, one that invoked a broader vision than in the past—a vision that looked inward toward Earth and its humanitarian problems as well as outward toward the planets and their intellectual challenges.

Increasingly aware of fading public support for the space program, and faced with the grim reality of a rapidly dwindling budget and what that implied for the future of JPL, William Pickering began to consider an alternative to space research for that portion of his unique technological resources not otherwise dedicated to supporting NASA's dwindling planetary exploration program. Perhaps the organization could support a research and development program in the civil sector as a contiguous, but entirely separate level of effort, from the traditional NASA programs that had characterized it in the past. But, as was to be expected, not everyone in the Caltech-JPL family agreed with Pickering's decision.

Endnotes

1 Hall, R. Cargill, *Lunar Impact: A History of Project Ranger* (Washington, DC: NASA SP-4210, 1977), p. 289.

2 Folder 120 in the William H. Pickering Speech Collection. Pasadena, California: Archives and Records Center, Jet Propulsion Laboratory, JPL-181, 2004.

3 Swaim, Dave, *Star-News*, Pasadena, California, 20 February 1965.

4 Ibid.

5 LeFigaro, 5 March 1965.

6 France-Soir, Vendredi, 5 March 1965.

7 L'Aurore, Vendredi, 5 March 1965.

8 Folder 122 in the William H. Pickering Speech Collection, Pasadena, California: Archives and Records Center, Jet Propulsion Laboratory, JPL-181, 2004.

9 Mudgway, Douglas J., Oral History Interview with William H. Pickering, Part 6D, Pasadena, California, July 2003.

10 Hall, R. Cargill, *Lunar Impact: A History of Project Ranger* (Washington, DC: NASA SP-4210, 1977), p. 301.

11 Ibid, p. 305.

12 Mudgway, Douglas J., Oral History Interview with William H. Pickering, Part 6B. Pasadena, California, July 2003.

13 The precise rate was eight and one third bits per second.

14 Sullivan, Walter, *New York Times*, 15 July 1965.

15 *New York Times*, p. 22, 15 July 1965.

16 *Time*, 23 July 1965.

17 Ibid.

18 James, Jack N., et al., "Mariner IV Mission to Mars." Pasadena, California: Jet Propulsion Laboratory, Technical Report No. 32-782, 15 September 1965.

19 Folder 134, 137 in the William H. Pickering Speech Collection, Pasadena, California: Archives and Records Center, Jet Propulsion Laboratory, JPL-181, 2004.

20 Folders 212, 217 in the William H. Pickering Office File Collection, Pasadena, California. Archives and Records Center, Jet Propulsion Laboratory, JPL-186, 2004.

21 Koppes, Clayton R., *JPL and the American Space Program: A History of the Jet Propulsion Laboratory* (New Haven: Yale University Press, 1982), p. 174.

22 Ibid, p. 179.

23 Mudgway, Douglas J., Oral History Interview with William H. Pickering, Part 6E. July 2003.

24 Swanson, Glen E., *Before This Decade Is Out: Personal Reflections on the Apollo Program* (Washington, DC: NASA SP-4233, 1999), p. 300.

25 Burrows, William E., *This New Ocean* (New York: Random House, 1998), p. 393.

26 Koppes, Clayton R., *JPL and the American Space Program: A History of the Jet Propulsion Laboratory* (New Haven: Yale University Press, 1982), p. 207.

27 Ibid, p. 208.

28 Newell, Homer E., "Surveyor: Candid Camera on the Moon." *National Geographic*, October, 1966.

29 Koppes, Clayton R., *JPL and the American Space Program: A History of the Jet Propulsion Laboratory* (New Haven: Yale University Press, 1982), p. 210.

30 Ibid, p. 210.

31 Ibid, p. 212.

32 Ibid, p. 212.

33 Folders 212-247 in the William H. Pickering Office File Collection, Pasadena, California: Archives and Records Center, Jet Propulsion Laboratory, JPL-186, 2004.

34 Folder 196 in the William H. Pickering Speech Collection, Pasadena, California: Archives and Records Center, Jet Propulsion Laboratory, JPL-181, 2004.

35 An astronomical term used to distinguish this situation from "superior conjunction" when Earth and Venus would be on opposite sides of the Sun and the distance between Earth and Venus would be at a maximum.

36 Wilford, John Noble, "Mariner 5 Passes Venus and Finds Magnetic traces." *New York Times*, 20 October 1967.

37 Mezitt, Beth Pickering, Private correspondence with the author, March 2005.

38 Sullivan, Walter, "Soviet Venus Data Pose Puzzles for U.S. Experts." *New York Times*, 20 October 1967.

39 Mudgway, Douglas J., Oral History Interview with William H. Pickering, Part 6E, July 2003.

40 Folders 105-107 in the William H. Pickering Office File Collection, Pasadena, California: Archives and Records Center, Jet Propulsion Laboratory, JPL-186, 2004.

41 For more information see Mudgway, 2005.

42 Sullivan, Walter, "Mariner 6 Gets Data Near Mars." *New York Times*, 31 July 1969 and "Mariner 7 Sends Sharpest Mars Pictures," *New York Times*, 6 August 1969.

43 Sullivan, Walter, "Huge Dish Antenna in Desert Catches Mars Photos." *New York Times*, Monday, 4 August 1969.

44 Folder 202 in the William H. Pickering Speech Collection, Pasadena, California: Archives and Records Center, Jet Propulsion Laboratory, JPL-181, 2004.

45 Folder 202 in the William H. Pickering Speech Collection, Pasadena, California: Archives and Records Center, Jet Propulsion Laboratory, JPL-181, 2004.

Chapter 8

New Initiatives

The institutional turbulence that clouded JPL relations with Caltech through the 1960s took a new turn in 1969 when Lee DuBridge, Caltech's long time president, mentor, and staunch supporter of William Pickering and his management of JPL, moved away from the Institute to become science advisor to President Nixon. DuBridge had always maintained a strong belief in the mutual benefits of the JPL-Caltech partnership, despite NASA criticism of JPL's management practices and his own faculty's growing doubt of the continued value of Caltech's association with JPL.

With DuBridge gone, Pickering would have to start over to build a viable working relationship with the new Caltech president who, like Pickering himself, was a physicist turned top-level administrator.[1] Former Secretary of the U.S. Air Force Harold Brown brought a background in the management of complex relationships between big science and national security to his new position. Early in his tenure, Brown initiated yet another committee to look into JPL-Caltech relations. Chaired by Professor Norman Brooks of Caltech, the committee decided that despite strong arguments to the contrary from the Caltech campus, the JPL-Caltech partnership should continue. However, it was critical of Pickering's civil systems projects, perceiving them as "unfocused and taking the Laboratory into areas in which it had little expertise." It saw the need for more "social science" in JPL's field of experience and more campus involvement in such nonspace programs. Some administrative changes were made, but in the end the Brooks report did little to change the status quo at the Laboratory and Pickering continued to run his civil systems programs much as he had done before.[2]

When asked why he embarked on a civil systems program in the face of the receding NASA budget rather than cutting back on staff levels, Pickering replied:

My answer was that I had a bunch of good guys working there and . . . where would I start cutting back and why? Also, if I had suggested much cutting back I would have had to argue with NASA. Another thing is that an organization of that size gets a lot of momentum behind it and it becomes an entity. You either keep it going or it disappears.

As for Caltech participation in JPL planning for non-classified and civil system programs, Pickering thought:

There should have been a lot more [JPL] contact with the faculty than there was. The reason [that] there was not, was due to the difference in cultures. The faculty had the individual entrepreneur type of culture while JPL had the team engineering, schedule, and cost type of culture, and those don't work well together.[3]

In October 1969, about a year after his "Science and the Urban Crisis" speech, William Pickering addressed the Los Angeles Philanthropic Association on the topic of "Space and the Humanitarian."[4]

Since earliest times, humans had learned with great success to exploit the resources of Earth. But they had not learned how to "equitably distribute those resources so as to minimize human deprivation and poverty . . . Our social progress has not kept pace with our spectacular material accomplishments. This lag in social progress is the concern of the philanthropist," Pickering said.

How then, Pickering asked "Do we relate space technology to the problems of the ghetto and the minority groups who are seeking greater social participation?" Pickering believed that our social problems "may have remained unresolved because of serious and deep disagreement about the goals and priorities of our national and local welfare and educational programs." Pickering suggested that the space program might be able to help find solutions to these types of social problems by "contributing experience gathered in the planning and management of some of the most massive technical programs in history." Systems analysis techniques, for example, could be used to improve understanding of the interactions between component parts of an overall social system. Adjustments could then be made in such a way as to attain the desired level of performance.

Such a process would require the merging of the broad disciplinary talents of scientists, engineers, and managers. "If sociologists, economists, politicians, and others of the behavioral and physical sciences are integrated into the system team, a structured approach can be made on wide social front," he said. Pickering cited successful examples of the application of these techniques in municipal fire and police departments, school expansion and vandalism problems, transportation, and air-pollution control.

While cautioning against too-rapid technical development that could give rise to secondary problems, Pickering called for a strong systems-based effort to devise a more equitable distribution of our resources that would extend our growing affluence to all segments of our society.

The following month Pickering traveled to London, England, where he delivered addresses to the Royal Astronomical Society and the British Interplanetary Society on the recent Mariner 6 and 7 missions to Mars. Large audiences of avid space enthusiasts supported his lectures on both occasions. Afterwards, the Interplanetary Society presented him with a special memento of the occasion as a mark of respect and admiration. It was a quick trip and was followed a few days later by another visit to New Zealand to address the Auckland Institute and Museum, an institution representing the local branch of the Royal Society of New Zealand. There he spoke in general terms of the Society's place in the Space Age.

What had been a busy year for William Pickering ended on a slightly ironic note when the Los Angeles Philanthropic Association elected William Pickering as its choice for "Outstanding American for 1969" and marked the occasion with the presentation of a splendid gold cup. That William Pickering, still an avid New Zealander after 40 years in this country, should be regarded as an "outstanding American" is a measure of the wide respect and esteem with which his public figure was held at the time.

In retrospect, 1969 seemed to mark a watershed in Pickering's total involvement in the space program. At this point, the Laboratory had survived and was running well under his direction. It had effectively seized the lead in the race to space, a goal that had engaged most of his attention in the early years, and had accumulated significant demonstrable evidence of its success in the field of planetary exploration. As evidenced by his public discourses, his attention now broadened to include "inner space"—the transfer of new technologies and system engineering principles (hitherto devoted to space initiatives) to the problems of social intercourse, communication, and human well-being. These new initiatives did not of course act to the detriment of on-going space projects, but allowed him to maintain the Laboratory workforce in the face of diminishing NASA budgets while at the same time demonstrating useful and practical examples of the transfer of space technology to the public sector.

One might also speculate that Pickering looked to the future, five or six years hence, when he would of necessity retire from JPL at age 65 and be in need of projects to which he could devote his talents for the remainder of his productive life. Perhaps the field of "civil systems" would provide a rewarding and challenging domain for such an enterprise.

Pickering was under no illusion as to the problems associated with successfully transferring the esoteric technologies that had grown out of the space program to the more mundane but no less important problems of urban management, as evidenced by a paper on "Practical Considerations of Technology Transfer," that he presented to the American Astronautical Society in March 1972.[5]

Pickering perceived that the public had begun to wonder when the "feats of high technological adventure" that they had witnessed and paid for in the past would produce some obvious practical benefits in the scenarios of everyday life such as "the seething ghetto, the ailing economy, the troubled social environment." Despite the enormous body of new technology and management "know-how" that the government had accumulated over the past several decades, there had been no progress in the social scene, education, or economics that could match the advances made in knowledge of our space environment. Noting that NASA had gone to great lengths to make its activities and scientific data generally available he said, "the dissemination of technical information is not identical with technology transfer. . . . The broad dispersal throughout society of space-generated knowledge is not the same as marketing a technology to fit, or match, a well-defined need in the civil sector."

The technology had not been transferred from the government sector to the civil sector, Pickering asserted, because industry had not been enthusiastic about accepting "government-generated information" and because the technologists had not adapted to a "marketing posture." He believed that the three groups involved (government, industry, and the civic sector) must learn to understand each other's environments and approaches to pressing problems. Pickering said it would be necessary to employ the concept of systems management to disperse the new technologies on a broad basis across many interactive user groups. He gave examples of how JPL was applying these principles to several civil programs in which it was engaged—one in city government, the other in the field of urban transportation.

But to achieve permanent success in technology transfer Pickering believed that "we must somehow deeply involve the social scientist in the application of technology. We must learn to satisfy the human condition with technological means while coping with the hard-nosed realities of modern life."

The new perspectives that Pickering had advocated in the late 1960s began to materialize at JPL in the early 1970s. A civil systems program office had been established under the management of Dan Schneiderman and, by 1972, it was actively pursuing a variety of nonspace-related projects directed to finding working solutions to current problems in the civil sector.

The *Los Angeles Times* reflected this new face of JPL in a 1972 article titled "JPL: Building a Better Mouse Trap is its Goal."[6] It was a far cry from its earlier headlines depicting the awesome achievements of JPL's exotic space-voyaging machines. This was really down-to-earth material, but it was realistic and reflected Pickering's broadened view of JPL's mission. The projects in work at the time covered a wide range of disciplines and applications such as transportation, the environment, law enforcement, education, and biomedical engineering.

The largest and perhaps the most complex of the JPL civil projects was the People Mover—an experimental computer-controlled transportation system then being constructed on the campus at the University of West Virginia at Morgantown. The Personal Rapid Transit (PRT) project, as it was officially called, represented most of the issues that Pickering had addressed in his public statements. Funded by the Urban Mass Transportation Administration, a government agency; designed with Space Age technology and systems management by JPL; built by Boeing, a leader in the aerospace industry; and sponsored by the University of West Virginia—the PRT project appeared to be assured of a successful outcome. By linking all three of the university's campuses with a computer-controlled fleet of up to 100 small passenger-vehicles moving over a network of elevated guideways, the system was intended to ease the students' problem of inter-campus commuting. Scheduled for completion in 1973, the PRT was viewed as the prototype for the development of similar systems in other areas of the country.

It should have been a success, but Pickering's hopes for a successful project were dashed when the project ran years over schedule and costs to complete the project far exceeded original estimates. Frustrated by precisely those factors that he had warned about, Pickering pulled the Laboratory out of the PRT once the initial development, for which JPL was obligated, was underway. Announcing the decision in his second "State of the Laboratory" message, Pickering told his staff "the Morgantown Project is being scrapped. I think that we can be very thankful that we got kicked out of that program. We got kicked out of it for the right reasons, because they weren't going to do the program correctly . . . and so we came to a parting of the ways."[7]

Eventually the PRT project was completed successfully without JPL's participation.[8]

In other projects, Schneiderman's scientists and engineers adapted the Laboratory's wide range of expertise to civil sector problems in the fields of air pollution abatement, law enforcement, education, and biomedical technology.

In all of these projects, Pickering insisted that the Laboratory's role was to work with sponsors to conceive and develop applications and to then make them available on a nonselective basis for manufacture by industry.

Pickering's civil systems program included another major project known as the Four Cities Program.[9] Begun in 1971, it was intended to explore and demonstrate new ways of linking federally-sponsored new technology with local government. It was implemented by assigning a science and technology advisor to the staff of the city manager for each of the four major California cities in the program. The advisors were drawn from major aerospace contractors in the region. JPL coordinated the program, evaluated the results, and provided technical guidance and support where necessary. Each science and technology advisor was given a mandate to study the technology transfer process, familiarize the city government with new technologies relevant to its problem areas, and look for market opportunities for their company. Feedback from these experiences in local government would, it was thought, benefit the aerospace companies by affording them greater awareness of social issues within their fields of interest.

At the end of the two-year period the evaluation report was optimistic, but fell far short of demonstrating the viability of the technology transfer process that Pickering was advocating.

Disappointing though the outcome of the Morgantown and Four Cities programs may have been, Pickering believed he understood the reasons for the adverse results and continued to advocate his ideas for applying the unique resources of his Laboratory to other problems of modern society.

Pickering's speeches in 1973 addressed those issues rather than more direct space-related topics that had characterized his discourses over the past decade. On two occasions he addressed branches of the IEEE; one in Boston, Massachusetts, the other in the San Gabriel Valley, California. Under the title "Reflections of the 1960s" he told the Boston Massachusetts branch of the IEEE that "in the future, the technologist may be dealing more with social problems—those of urban sprawl transportation, the environment, troubles of the people. He will be confronting the quality of life more than the reliability of an electronic circuit. Society will be his customer more than government." He went on to characterize the response of society as more Darwinian (that is, more evolutionary) than the Newtonian (or deterministic) environment that they were used to. The engineer and scientist of the future will have to deal with the "formless patterns of politics, economics, sociology, and psychology," he said. Warning that it would nevertheless be difficult to adapt to these changing motifs, he said that it would become more and more necessary in the decades to follow.[10] Similar themes ran through his address on "Technology in the Waning Century" to the San Gabriel Valley Branch of the IEEE.

Pickering's ideas, opinions, and experiences with technology transfer received an airing at the highest level when he testified before the Senate Commerce Committee's Subcommittee on Science, Technology, and Commerce in September 1973.[11] There, Pickering traced the growth of JPL's interest in transferring its intrinsic skills in aerospace technology to the civil sector and described instances where JPL had found productive areas for the application of those specific technologies that we saw earlier: medical engineering, the Four Cities program, transportation, etc.

Based on his experience in dealing with problems in the public sector, Pickering offered the following observations. First, technology transfer is a slow process—it takes a long time to understand the social, legal, and economic aspects of a technological solution to a civil sector problem and existing funding was not commensurate with that fact—and in the definition process there is a diffusion of responsibility and decision-making authority (for introducing new technology into existing systems). Also, there is distrust between public officials and technologists (regarding the advantages of introducing new technology into existing systems) and no mechanism exists to make civil sector technology that had been developed in federally-funded laboratories available to industry and commerce.

He offered similar comments with regard to federal programs of technology utilization and transfer with an additional, rather scathing criticism. "Most federal sponsors outside of NASA, DOD, and AEC either do not understand the research nature of a technical problem or they do not appreciate the problems of application. However, NASA and DOD also do not understand the social content of many national problems," he said.

Typical of Pickering, it was an intense, credible, hard-hitting elucidation of the subject based on his unique, up-to-date experience with real-life situations. Among those present, there was simply no one to challenge either his facts or the conclusions he drew from them. He obviously caught their interest for he was called upon by one of the Senators to elucidate further on his testimony relative to establishing a network of Regional Technology Applications Centers.[12]

By 1974, biomedical projects had become the largest part of the JPL civil systems program. In his annual "State of the Laboratory"[13] address in 1974 William Pickering talked of "the possibility of forming a medical sciences laboratory at JPL . . . with participation of the campus and some of the research hospitals in the area." The purpose, as Pickering explained, was to "exploit the science capability at campus, the engineering capability at the Laboratory and the medical research at some of the hospitals." Funding would be partly private and partly government and when it reached maturity the new medical laboratory would be "spun off." Feasibility studies were underway he said and both he and the president of Caltech believed "it is a good thing to do and we are hoping it will come about."[14]

Early in 1971, the Caltech sponsored a lecture series on "Systems Concepts: Contemporary Approaches to Systems." At this series, William Pickering presented a paper titled "System Engineering at the Jet Propulsion Laboratory" in which he traced the evolution of JPL to a systems engineering organization from its earliest times with the U.S. Army contracts to the present time. He explained how the Sergeant program represented a classical systems engineering task and how the experience gained therein migrated to JPL's lunar and planetary projects after it transferred to NASA in 1958.

Systems engineering required the optimization of the overall end-to-end system rather than the suboptimization of the individual elements of the system. Such a process was accomplished in a sequence of discrete steps beginning with a clear definition of the ultimate objectives and ending with an implementation plan. For lunar and planetary projects—which involve a spacecraft system, a tracking and data system, and a mission operations system—these steps became of critical importance and were followed meticulously in each case. He gave details of how each system was broken down into its functional elements to eventually reach a level of complexity commensurate with a single element. The functional boundaries of individual elements were established and controlled by defined interfaces.

Designing for success, testing for performance, mission planning, and project management were the key factors in carrying out a successful space mission and Pickering explained each of them in considerable detail.

Pickering viewed the practice of engineering as more concerned with management, information, and good judgment rather than with mathematical analysis. Wherever possible mathematical techniques were used to solve optimization problems, but many decisions had to be made with little or no quantitative information and for these cases good judgment and experience were required.

The Laboratory was now exploring the viability of a systems approach to solving problems in the civil sector and he gave examples in the areas of mass transportation and urban health systems.[15]

Important and newsworthy though it was at the time, the civil systems program generated little or no impact on the Laboratory's ongoing space programs and that was how Pickering intended it to be. He hoped to build up sponsorship for some of the civil systems programs to the point where they would be essentially self-supporting.

Meanwhile the planetary programs office, funded by NASA and directed by Pickering's former student Robert Parks, pressed on with the two planetary projects of immediate concern: a mission to Mars planned for the 1971 opportunity and another to Venus and Mercury planned for launch in 1973.

Space

The 1971 Mariner mission to Mars was NASA-JPL's most ambitious planetary project yet and the data it returned from Mars compelled scientists to reconsider long-held opinions of the origin and subsequent history of that intriguing planet.

The primary objectives of the 1971 Mars mission were to search for evidence of life and to gather data that would aid the design of a later Mars lander mission that would extend and intensify the search from selected locations on the Martian surface. To this end, each of the two spacecraft carried a comprehensive suite of scientific instruments that included a high-resolution television imaging system designed to photograph up to 70 percent of the entire Martian surface. Two of the country's most prominent planetary scientists, Carl Sagan of Cornell and Bruce Murray of Caltech, headed the Mariner Mars Imaging Team. The sophistication of the new spacecraft was matched by a complementary enhancement of the data collection and data processing capabilities of the worldwide deep space network (DSN) and the Space Flight Operations Facility at JPL.[16]

Now well experienced in the designing, fabricating, and testing of planetary spacecraft, Mariner teams prepared two identical spacecraft for launch during the Mars opportunity in May 1971. The first spacecraft, identified as Mariner 8, was lost when the Centaur second stage launch vehicle failed a few minutes after launch.

Acutely aware that the Soviets were also launching to Mars in this window of opportunity, Pickering agonized over loss of Mariner 8. He would have known that the Soviets had hastily prepared a Mars orbiter spacecraft which they called Kosmos 419, specifically to preempt NASA-JPL's Mariner attempt to be first into orbit around another planet.[17] He probably did not know that the Kosmos 419 had been launched from the Russian launch site at Baykonur the day following the Mariner 8 launch and that it too had experienced a failure during the launch sequence. A week or so later, when the Soviets launched two more Mars-bound spacecraft both heavy Mars landers and both injected successfully onto a Mars intercept trajectory, he would have been most apprehensive about the outcome of the Mariner 9 mission.

Pickering would have observed its successful launch and injection onto a near perfect trajectory to Mars on 30 May 1971, with a high degree of relief. As far as he could have known, based on meager progress reports coming from Russian sources, the Soviet Union then had three spacecraft en route for Mars—the U.S. had one. Mariner 9 thus represented Pickering's only opportunity to maintain the preeminent position that he had struggled so hard and for so long to achieve for the U.S. For Pickering the "space race" was still very much alive and he perceived the Soviet challenge as a threat that he was personally obliged to resist.[18]

By the time the spacecraft arrived at Mars in early November, an immense dust storm completely covered the planet, obscuring the vital Martian surface features that were a prime objective of the mission. According to observations from Earth-based telescopes it was the most severe dust storm ever recorded in terms of density, area, and persistence.

The first news reports were not hopeful. "Dust Storm Smears Mariner Photos of Mars," headlined the *Los Angeles Times*. The 5-mile high cloud of dense reddish-yellow dust appeared to be moving at about 20 to 30 miles per hour across the Martian surface and completely eliminated any possibility of photographing the surface features beneath. "While the storm may hinder mapping and photography of the surface features for a time, it will offer an excellent chance to study one of the great dynamic features of the planet," Pickering said.[19] The next day Mariner 9 again executed a flawless maneuver to slow down and enter a precisely-controlled orbit around Mars and in doing so became the first spacecraft to orbit a planet other than Earth.

By this time mission controllers and science teams working together had come up with a new strategy. Making use of the innovative reprogrammable computers carried by the spacecraft, they would defer the original mission plan that called for systematic mapping of the Martian surface to allow time for the dust storm to clear. Meanwhile the imaging and remote sensing instruments would be reprogrammed to allow scientists to conduct observations of the storm-related phenomena as the opportunities arose. Pickering quickly endorsed the plan and the necessary commands for in-flight reprogramming of the spacecraft computers were transmitted from the DSN to the distant spacecraft.

Meanwhile, the two Soviet spacecraft steadily approached their appointment with destiny at Mars. Back at the Soviet mission control center, Russian scientists and controllers must have watched with horror as JPL reports of the huge dust storm appeared regularly in the media around the world. They surely realized that the two Soviet spacecraft with their fully automated landing and science data gathering sequences would be extremely vulnerable to the deadly effects of the dust storm then pervading the Mars atmosphere and that there was nothing they could do about it. Their premonition proved to be justified. In what must have been a heart-wrenching event for the Soviets, Mars 2 suffered

Mariner 9 spacecraft (Photo: NASM Archive, Image 71-h-717).

a malfunction as it entered the Mars atmosphere for the final landing sequence on 27 November 1971 to become "the first human-made object to make contact with Mars."[20] Although "make contact with Mars" sounds rather like a prosaic way of saying "crashed on Mars," it nevertheless represented a very significant achievement at the time, even if it was not the successful outcome that its designers intended. However, the second spacecraft, Mars 3 did perform its preprogrammed landing sequence successfully. It touched down on the Martian surface on 2 December 1972 to become "the first human-made object to perform a survivable landing on Mars." Soviet hopes for further success were dashed however when the first TV transmission from the spacecraft, showing only a "gray background with no detail, abruptly ceased after 20 seconds."[21]

How Pickering would have empathized with those Soviet scientists, recalling his desperate days with the first Ranger missions when a failed electronics device doomed the entire mission to failure, despite its considerable achievements in reaching that point in the mission. It should be recorded, however, that "despite the failure of the Lander imaging system, the two Russian orbiters carried out a full cycle of scientific experiments until contact with both was lost in July 1972."[22]

In mid-January 1972, the Mars atmosphere began to clear and spacecraft controllers initiated the surface mapping sequences, periodically synchronizing the spacecraft orbit to coincide with the view periods of the giant new 64-meter diameter antenna at Goldstone. An avalanche of science and imaging data began to flow regularly from Mars to Goldstone and to the scientists at Pasadena. The astounding images and other science data afforded scientists an entirely new appreciation for the genesis and development of the planet. A giant volcano, Olympus Mons, that dwarfed anything seen on Earth; an enormous canyon almost as long as the U.S. is wide and over six kilometers deep in places; braided channels that looked as though they were formed by flowing water and resembled terrestrial river valleys; images of the two moons of Mars, Phobos and Deimos; and radio science studies of the composition of the Mars atmosphere were just part of the cornucopia of new science that Mariner 9 delivered to Earth in the ensuing months.

The spacecraft continued in full operation until late October 1972 when, with its gas supply depleted and its attitude stabilization system disabled, the spacecraft began a slow uncontrolled descent to eventual destruction in the Mars atmosphere. The Mariner 9 mission was over. It had fully met and even exceeded its mission objectives and survived more than a year beyond its original design life. To Pickering, those facts indicated that his Laboratory had finally ascended the learning curve. As he always knew it would, it had learned how to build planetary spacecraft. As far as Mars exploration was concerned, he was satisfied that the U.S. had reasserted its premier position. For him personally that perception was very satisfying indeed.[23]

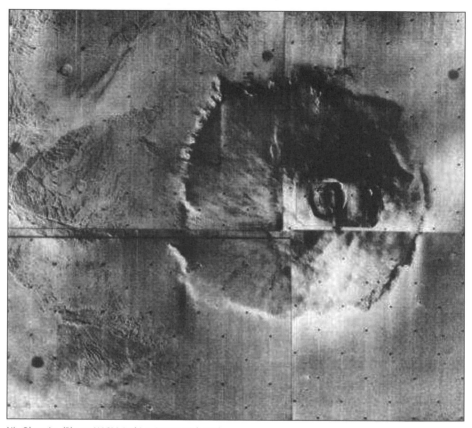

Nix Olympica (Photo: NASM Archive, Image 72-h-141).

Earlier that year he had prepared an article for a local newspaper titled "Odyssey to the Rim of the Solar System," in which he discussed Mariner 9 and its objectives and talked of future missions to Venus, Mercury, and Mars with landings on the Martian surface to search for evidence of primitive life. He also envisioned long duration missions that would use a new technique called gravity-assist to leapfrog from one planet to the next and thereby enable a space-craft to eventually reach the very edge of the solar system and even beyond. Such a mission, already under study at JPL, was called the Grand Tour.

He predicted:

> Missions to the planets, and particularly to Jupiter, will tell us a great deal about the origin and evolution of our solar system and the inception and development of the life forms we know on Earth. During the remainder of this century, the thrust of man's probing intelligence will reach out to the very edges of the solar system.[24]

By April 1972 when he delivered "Thirty Trillion Bits" to a combined meeting of the AIAA/IEEE/SAMPE[25] in Salt Lake City he had a wealth of impressive data on the Mariner mission to share with his audience. Mariner 9 traveled approximately 247 million miles in 168 days to arrive at Mars on 13 November within 38 miles of its aiming point and within 2 minutes of the estimated arrival time. After the dust cleared from the Mars atmosphere eyes, Mariner 9 mapped 90 percent of the Mars surface and transmitted at least 5,700 pictures at approximately 5,250,000 bits each. In thousands of photographs Mars was revealed as a dynamic evolving world with structural features never seen before from Earth—features that included indications of recent volcanic activity, erosion processes apparently related to once flowing water, and vast stretches of sinuous canyons. Data returned from the other scientific instruments would enable significant revision of the scientific model of the planet and ranging and motion studies would aid in refining the geometry of Mars celestial mechanics environment.

"The Mariner 9 mission," he said, "is expected to return 12 to 15 times as much scientific data as all previous planetary missions combined."

Looking to the future, Pickering said that "although Mariner data is adding intriguing new dimensions to our knowledge of Mars, it is extremely unlikely that the existence of life forms on the planet can be determined from orbit. That demonstration must wait such missions as the 1975–1976 Viking project to orbit and land two spacecraft on the surface of the planet."[26]

In March of 1973, Pickering initiated what was to become a regular event—the Director's annual "State of the Laboratory" address to the entire Laboratory staff. In this, the first of its type, he reflected the austerity of the times and the impact on the Laboratory's activities caused by the hiatus in NASA's planetary program. He spoke of receding budgets, diminishing staff levels, relations with NASA and Caltech and industry, and the vulnerability of JPL to criticism and even serious cutbacks. While he saw a prospect for the nation to "continue a civilian space program of considerable size," but he warned that "the era of technological laissez-faire was at an end."

Concerned that in some areas JPL had acquired a less than desirable reputation as an expensive organization with which to do business he said, "we must make a strong effort to correct the gold-plating image that JPL has unfortunately acquired," and cited the current, fixed-price Mariner Venus-Mercury project as a significant step in that direction. He saw good prospects for the future of non-NASA projects: biomedical engineering, energy, environment and transportation, and public safety were promising fields. He concluded by emphasizing the need to consider costs at every level in the Laboratory's activities. "The future is not exactly ready and waiting as it was in 1958 . . . The problems are there but the money is harder to find. This time we must work to get it."[27]

In terms of NASA funding, the 1973 Mariner mission to Mercury and Venus epitomized the austerity of the 1970s and Pickering's concern with cutting costs. If the Laboratory was to survive in this environment, it was imperative that the costs to NASA of new planetary missions be reduced to an absolute minimum. That Mariner 10 was so unequivocally successful, both financially and scientifically, was a testament to the ingenuity, skill, motivation, and experience of the NASA-JPL teams that brought it to fruition and the science teams, most notably the imaging team led by eminent planetary scientist Bruce Murray of the Caltech, that gathered the science data and interpreted it.[28] It was at once a product of the call to "do better with less" that Pickering had issued in his first "State of the Laboratory" address and a demonstration of how the Laboratory could respond to NASA's straightened circumstances to keep the planetary exploration initiative alive and productive.

Launched at the end of 1973, Mariner 10 reached Mercury in March 1974 after a close flyby of Venus during which it made important science observations including imaging of that planet and made use of Venusian gravity to assist its flight to Mercury. It made three flybys of Mercury collecting, among other scientific data, some 2,300 detailed images that showed the surface of Mercury to be more Moon-like than Earth-like. The science data from Mariner 10 exceeded the scientists' greatest expectations.

But Mariner 10 could boast more than its scientific accomplishments—it also represented a number of major advances in deep space technology. It demonstrated, for the first time, the exquisitely complex technique of gravity assist: using the gravitational field of one planet to modify the trajectory of a spacecraft to enable it to reach another. Making use of all three giant 64-meter diameters of the DSN it returned data from Mercury at the unprecedented data rate of 118 kilobits per second and, using dual frequency radio transmissions from the spacecraft to Earth, it enabled radio science observations and demonstrated the improved efficiency of higher frequency radio communications from deep space. Finally, it was the first spacecraft to perform multiple encounters with a target planet.[29]

Long and careful planning enabled the project to meet its cost and performance goals and the cost-plus-fixed-fee contract with the Boeing Company that built the spacecraft showed that the Laboratory could now successfully manage a large, schedule-intensive contract with industry. In an overall sense, Mariner 10 represented the culmination of more than a decade of the Laboratory's evolving expertise in all aspects of the technology required to execute a successful planetary mission in deep space. But it was also the last of the Mariners—and the last of an era.[30] Planetary missions of the future would be bigger, more costly, more complex, and their management and direction would no longer be carried out by the close-knit group of brilliant, egocentric experts that William Pickering had assembled, inspired, nurtured, and defended throughout his years at JPL.

Pickering remained intensely interested in future planetary missions, but now he increasingly left the public presentations and recognition of the Laboratory's space program to others and devoted more and more of his personal attention to the issues of technology transfer and the problems of Earth rather than those of space.[31]

Society

The depth of William Pickering's evident concern for the well-being of society as a whole—perhaps more correctly, civilization as a whole, since most of what captured his interest pertained to the most advanced societal groups—is exemplified by a list of his public addresses in the last two years of his tenure at JPL. Among list of titles:

January 1974: New Challenges for the Engineer: "The urgent need is to improve the efficiency of existing energy-consuming processes and to develop new techniques that will make the nation essentially independent of international politics and able to pursue uninterrupted programs of orderly social development."[32]

January 1974: Improving the Environment: "It is mandatory that we recognize the urgent need to commit all of the high technical skills developed for aerospace applications into improvement of the environment."[33]

May 1974: Understanding the Universe: "Modern civilized man's only hope may be to learn to understand his universe [and to] create a society flexible enough to cope with an environment that changes overnight."[34]

May 1974: Digitizing the Social Conscience: "We must look to science and engineering to nurture a renaissance for our society if we are to enter the new century with clear prospects for survival."[35]

May 1974: Toward a New Society: "Science and technology will be needed even more than in this century or else the quality of life and our society will sharply decline."[36]

June 1974: Inheriting the Future: "Cybernetics alone cannot ensure survival . . . Your ultimate challenge will be to control, even eliminate the threat of growing scarcity of raw materials and the increasing rate of consumption by a burgeoning population."[37]

November 1974: The Next One Hundred Years: "The grim irony is that the gap between the rich and the poor nations is continually widening, despite our growing technological and communications capabilities . . . The evidence is increasingly clear that, in the next one hundred years, we must progress from selfish independence to cooperative interdependence, a condition where we substitute the regional blindness of nations for the overwhelming good of the race."[38]

In June 1974, Pickering prepared a proposal for submission to the Committee for Space Research (COSPAR), a prestigious international body dedicated to promote, on an international level, scientific research in space with emphasis on the exchange of results, information, and opinions. The proposal called for the establishment of "An International Solar System Decade" that would run from July 1976 through June 1986.

Just as the First Polar Year of 1882, the second Polar Year of 1932, and the International Geophysical Year of 1954 had reflected the scientific interests and observational capabilities of their times so, said Pickering, the International Solar System Decade (ISSD) could make use of presently available rockets, spacecraft, and advanced data acquisition techniques to support an international effort "directed at a better understanding of our whole solar system." In December 1974, in letters to Roald Z. Sagdeev of the Soviet Academy of Sciences and to G. Contopoulos, General Secretary of the International Astronomical Union, Pickering sought their support for COSPAR sponsorship of his proposal but to no avail.[40] "That [idea] never went very far—it was an example of my naiveté about political motives." But the idea was not dead and he did get another opportunity to pursue the topic of international cooperation in space.[41]

As 1975, Pickering's final year of tenure at JPL, wound down, many of the non-NASA organizations with which he had been associated during his long career recognized his contribution to their various areas of interest with nominations for their highest honors, or awards. The Collier Trophy, L. M. Ericsson Award, Marconi National Fellowship, George C. Marshall Medal, Delmer S. Fahrney Medal from the Franklin Institute, the WEMA Medal of Achievement, and Advancement of Engineering Award from University of Southern California were examples in the fields of aeronautics, astronautics, or aerospace technology. On each occasion, his acceptance speeches focused on themes of society and technology—topics that dominated his thinking at this period.

His contribution to science and regard for his greatly admired public figure in New Zealand had not gone unnoticed either. On 26 November 1975 he was immensely gratified with the notification that Her Majesty Queen Elizabeth II of England had graciously invested him as Honorary Knight Commander of the Most Excellent Order of the British Empire. The prestigious distinction recognized his "Contributions to Science" and carried the abbreviated nominals K.B.E. The "Honorary" prefix signified his formal, non-British citizenship. By tradition, the title would be conferred on the Queen's birthday in June the following year by her representative in New Zealand.[42]

Toward the Future

Two major new NASA initiatives, Viking and Voyager, dominated the closing years of Pickering's tenure at the Laboratory. Both had a long period of gestation within the NASA organization, but by 1975 both had been defined, approved, funded, and assigned to NASA centers for implementation. Viking, a major project designed primarily to search for biological evidence of life on the surface of Mars would be managed by the Langley Research Center in Hampton, Virginia. Voyager, a derivative of an earlier concept known as the Grand Tour, was an ambitious attempt to explore several of the outer planets in sequence, beginning with Jupiter, and would be managed by JPL.

The Viking launches, there would be two of them, were planned for the August-September Mars opportunity in 1975, while the two Voyagers were scheduled for the August—September opportunities in 1977. Both missions would depend upon JPL's DSN for their tracking and data acquisition support.

While his Flight Project Managers began to implement the daunting tasks of bringing these awesome new missions to reality at JPL, Pickering began to introduce related issues to his public audiences. Startling titles such as "Homo Sapiens; One of a Kind?," "Is there Life on Mars?," and "Extra-terrestrial Life: The Search Begins" were typical of his speeches in this period. In these talks, which attracted considerable public attention, Pickering used slides from recent Mariner missions and his redoubtable knowledge of physics, chemistry, and astronomy to make a case for justifying a search for elementary life forms on Mars. "If you accept the evidence that more favorable conditions might have existed on Mars in the past, it is altogether possible that life forms might have flourished at that time, and later learned to adapt to the present austere environment. . . . Considering the amazing variety and durability of life forms found on Earth, it seems improbable that we will detect none on Mars. It may be elusive and exist in extremely subtle variations but, if it is there, we should be able to find it." He followed these intriguing ideas with a description of the Viking Mars mission and told of the ways in which it would, hopefully, help to resolve questions that they engendered.

Looking beyond our solar system, he said, "The search for extraterrestrial life and advanced civilizations has largely been centered on radio astronomy techniques because of the enormous distances involved." He gave credence to the continuation of current programs to pursue the search, by citing new discoveries in the field of cosmological physics and chemistry.

Recent studies of intergalactic processes, he suggested, showed that "it is not unreasonable to presume that, given similar circumstances, the same kind of events that led to the origin of life on Earth are also occurring elsewhere in the universe. Astronomers are showing us that molecules identified in interstellar space are the progenitors of the protein and nucleic acids that are the basis to Earth life."

Turning to the philosophical implications of his visions he observed that "if one day we find traces of life on Mars, Titan, or in the atmosphere of Jupiter, the implications to human civilization will be enormous. Philosophy, religion, science, technology—all of the human arts—will have to be recast in a new image. The old ways will be suspect more than ever before. Man must then confront his dilemma; he must acknowledge the unacknowledgeable—we are not alone in the universe. Somewhere out there, the inevitability of other advanced, perhaps far superior, civilizations would become manifest. "Perhaps," he surmised, "nature did not throw away the pattern when she conceived us where nothing had existed before."[43]

In this pervasive environment, the huge Viking and Voyager projects moved steadily forward, dominating the work structure and straining the Laboratory, while the four civil systems programs—energy and environment, biomedical engineering, transportation, and public safety—continued to claim a small but vital part of Pickering's interest and attention.

At the end of 1975, just before his 65th birthday, Pickering called his supervisory staff together in the large new auditorium of the La Canada Intermediate School to hear what would be his final "State of Laboratory" message. By then it was generally known that he would be retiring within the next month or two and, in fact, plans were already being made to mark the occasion with appropriate farewell functions.

If Pickering harbored any feelings of nostalgia or regret on this occasion, they were not apparent in his very upbeat address. It was all "business as usual" for now. "Today I am not going to talk about retirement; that's several months off and I'll worry about it later," he began. He spoke of recent NASA action that confirmed JPL's primary role in planetary and lunar unpiloted missions and how that fact assured the future of the Laboratory. Current NASA programs that included Viking and Voyager were making good progress. He touched on JPL involvement in some automated Earth orbital missions and JPL's future interest in an active energy program for the Energy Research and Development Agency (ERDA). The work in civil systems had expanded to include a waste-water treatment plant and the Department of the Interior was looking to JPL for help with new technology applications to coal mining. This non-NASA activity continued to cushion the effect of inevitable reductions in NASA programs, he said.

He concluded on a high note:

> The Laboratory is in good shape with NASA . . . it has an opportunity to grow its activities with ERDA and other government agencies. If we continue to maintain an excellent staff, our opportunities for the future are just as good as they have been in the past. They are oriented in a somewhat different fashion than they were 10 years ago. But they are there and it is up to us to take advantage of them.

These were to be his last words as Director to the people with whom he had come so far, and whose standards of performance, dedication, and work ethic he valued so much and had defended so strongly. The essential elements required to carry the Laboratory toward the future were all there—it would be up to others to nurture and sustain them.

William Pickering turned 65 on 24 December 1975 and, in doing so, became subject to the Caltech institutional limitations on the maximum age for top-level administrators. Since the Caltech effort to find a successor had not yet run its course, Pickering agreed to remain in office a further three months, until 31 March the following year, to allow that to be completed.

Pickering regarded the situation with some equanimity. He recalled:

> So I had to retire . . . but having been around the lab for more than 20 years or so, I was beginning to feel myself, that it was time [for me] to go, and [for Caltech] to bring in new blood. Once the general concept had been established that I was going to leave shortly, the question of who would succeed me came in. . . . There was never any serious consideration of picking someone from the lab to succeed me, but rather getting someone from outside, particularly someone with a technical administration, governmental-type of background . . . that was the concept.[44]

JPL planned to bid him farewell with two formal social events in March 1976. These events were to be followed a few days later by a farewell reception at NASA Headquarters in Washington, DC. The AIAA also planned to hold a special "Thank You Bill Pickering" function, a few weeks later still in Los Angeles on 23 April.[45] The next few months it appeared would be a busy time for both Muriel and William.

As the prospect of his retirement from JPL and the NASA space program moved ever closer to reality, Pickering considered what he might do with the rest of his life. At 65 he was at the peak of his powers, physically fit, and mentally vigorous. His outstanding technological acumen and his unique achievements in space exploration had brought him worldwide recognition. His distinctions, awards, and honors were legion. His enviable record offered an open door to many options in industry, government, and academia. But, considering his roots, he inevitably felt drawn back to academia and the long-standing agreement he had made 20 years earlier with Lee DuBridge, former President of Caltech. DuBridge was long gone from Caltech by then, but Pickering held no doubts that the incumbent president, Harold Brown, would honor the agreement and allow him to return to Caltech as a full professor of electrical engineering if he so desired.

Pickering recalled the circumstances:

> When I had been active in the department in the 1940s . . . electrical engineering meant big motors, transformers, and generators.

So I go back 20 years later, and what do I find? There is no power engineering of any sort—everything is electronics or computers . . . So I thought about it a bit and thought that maybe I could work up a course that is a little bit on the fringes . . . and just about that time this guy from Saudi Arabia showed up!"[46]

Caught up in a whirl of social events related to the impending retirement, Muriel and William found little time to contemplate the impact to their well-ordered lifestyle that severance of William's connection to JPL would bring about. In Pickering's view, however, it was to be "retirement from JPL" and not in any sense retirement from an active and productive life, as future events would soon show.

The time passed quickly and soon enough the long-planned series of official retirement events began. On Friday, 19 March, several thousand employees, spouses, and guests gathered at Pasadena's spacious Convention Center to bid farewell to Dr. and Mrs. Pickering. The popularity of "Mr. JPL" buttons, worn with obvious delight by many of the attendees, served to heighten the political convention-like effect. Highlights of the evening included the unveiling of a magnificent near life-size portrait of William Pickering, a work by well-known artist Art Beeman who was also a long-time JPL employee.

NASA paid tribute to William Pickering the following week in a formal reception at NASA Headquarters in Washington, DC. The formalities included speeches and gifts (of which there were a considerable number) from the NASA executives and from the Center Directors. Among the gifts he received that evening were two of unique historical significance. One was the spare transmitter from the Explorer 1 Earth satellite from 1958 donated by the Kennedy Space Center; the other was the camera from the Surveyor 3 spacecraft that had soft-landed on the Moon in April 1967. Retrieved from the lunar surface and returned to Earth by Apollo 12 astronauts in November 1969, the camera had been the subject of an intense evaluation and found to be in excellent condition. Pickering was most impressed.[48]

Dr. and Mrs. Pickering with the near life-size portrait executed by Art Beeman of JPL, 19 March 1976 (Photo: JPL Photo number P16519B).

Employees To Honor Retiring 'Boss' At Gala March 19 Reception

The send-off will be nothing short of royal when JPL employees gather Friday night, March 19, in Pasadena's Convention Center to honor the Lab's retiring Director, Dr. William H. Pickering.

Dr. Pickering, Director of JPL since 1954, is ultimately retiring later this spring. He is being succeeded by Dr. Bruce Murray, whose appointment was announced last June by Caltech President Harold Brown.

A host of special activities and events are planned to mark the gala evening.

Highlighting some of the preliminary fun will be distribution of a specially designed "Mr. JPL" badge, featuring a caricature of Dr. Pickering.

Tickets, which include a badge, will be on sale at $1.50 each through JPL supervisors well in advance of the event.

The reception, with "no-host" bar facilities, runs from 6 to 8 p.m. Parking is available in the convention center's underground parking facility.

Complete details will be carried in the February 27 issue of UNIVERSE.

"Mr. JPL" button created by JPL caricaturist Bill Stephenson for the Pickering farewell event, March 1976 (Photo: JPL Archives, Universe, vol. 16, no. 14, 13 February 1976).

The event was a fitting tribute from the administration to one who, since long before most of its current members were in office, had played a major part in bringing NASA to its present preeminent position in humankind's efforts to explore the solar system.

Pickering's last formal appearance at JPL took place on 31 March 1976 at a ceremony before a full gathering of JPL employees on the beautiful central plaza now resplendent with a fountain, colorful flowering planters, mature trees, and many varieties of carefully tended shrubs. President Harold Brown of Caltech formally introduced Dr. Bruce Murray, well known to JPL scientists for his leading role on JPL's planetary imaging teams and former professor of planetary sciences at Caltech, as the new Director of Jet Propulsion Laboratory.[49]

It was over. After more than 20 years of leading the organization that he cherished almost as his own—which to some extent it was since he had, against all odds and opposition, made it in his own image—he had finally handed it over to someone else. Where would it go and how would it fare—well, only the future would tell but of this he was certain: he would not be a part of it.

That day, the *Los Angeles Times* paid an eloquent tribute to William Pickering. Writing that he "led the team whose talents and accomplishments brought the Moon, the planets, and the stars within man's reach." The *Times* continued, "Under Pickering's leadership, JPL was responsible for a record of formidable and almost incredible achievements in the unmanned exploration of space."

Recognizing that such magnificent accomplishments were always the result of team efforts, the *Times* writer noted the outstanding record of JPL in the era of space exploration and wrote "inseparably involved in that record is William Pickering, whose cool and quiet leadership has accounted for so much. His adopted country is grateful to this native of New Zealand for all that he has done. His career has been one of achievements that do honor to himself, his profession and his nation."[50]

Harold Brown, President of Caltech; William Pickering, retiring Director JPL; and Bruce Murray, incoming Director JPL, 31 March 1976 (Photo: JPL Photo number P16433A).

In the years after he left JPL as Director, Pickering was always made to feel welcome whenever he chose to visit the Laboratory. Security guards waved him through the main gate, he retained an assigned parking place, and he was made to feel like an honored guest by the entire staff. He frequently took visitors on tours of the Laboratory and was always on the invitation list for events of significance to its programs. These were the courtesies that meant most to him now.

If the fact that he no longer determined the course of action at JPL concerned him at all, it was not apparent either at the time, or in later years. "I just walked out," he said, and "never looked back."[51] Would he have gone back if he had been asked? "My answer to that is simply that I was never invited back. If I had been approached by someone at JPL [to return] for some reason, I'm sure I would have gone back," he answered.

Perhaps—but then at the time, he had other things on his mind.

Endnotes

1 See "Caltech's New President" in *The DuBridge Years*, Engineering and Science, California Institute of Technology, December 1968.

2 Koppes, Clayton R., *JPL and the American Space Program: A History of the Jet Propulsion Laboratory* (New Haven: Yale University Press, 1982), p. 239.

3 Pickering, W. H., Interview with the author. Pasadena, California, July 2003.

4 See Folder 217 in the William H. Pickering Speech Collection, Pasadena, California: Archives and Records Center, Jet Propulsion Laboratory, JPL-181, 2004.

5 Pickering, W. H., *Some Practical Considerations in Technology Transfer* (Springfield, Virginia: American Astronautical Society Publication 72-033, March 1972).

6 Austin, Lee, "JPL: Building a Better Mousetrap is Their Goal." *Los Angeles Times*, Sunday, 17 September 1972.

7 Pickering, W. H., "JPL State of the Laboratory Message." 18 April 1974 (WHP personal files, DJM copy with speeches).

8 See *http://web.preby.edu/~jtbell/transit/Morgantown/*. Accessed 7 July 2007.

9 Macomber, H. L., and James H. Wilson, "California Four Cities Program 1971-73." Pasadena, California: Jet Propulsion Laboratory, SP 43-4, May 1974.

10 Pickering, William H., "Reflections on the 1960s," Speech to the Awards Meeting of IEEE, Boston Branch. Newton, Massachusetts, Harold Wheelock Collection 1061-26, 11 March 1973.

11 Pickering, William H., "Statement Prepared for the Senate Commerce Committee, Subcommittee on Science, Technology and Commerce." Jet Propulsion Laboratory, Wheelock Collection 1061-36, 4 September 1973.

12 Pickering, William H., "Response to Senator Tunney: The Regional Technology Applications Center." Jet Propulsion Laboratory, Wheelock Collection 1061-89, 24 September 1973.

13 Pickering, W. H., "State of the Laboratory Message." 18 April 1974.

14 Although JPL continued to make important progress in biomedical applications, the idea of a medical laboratory went no further during the Pickering years at JPL.

15 See Chapter 7 in *Systems Concepts: Lectures on Contemporary Approaches to Systems*. Ralph F. Miles, ed. (New York: John Wiley & Sons, 1973).

16 Mudgway, Douglas J., *Uplink-Downlink: A History of the Deep Space Network* (Washington, DC: National Aeronautics and Space Administration Special Publication-4227, 2001).

17 Siddiqi, Asif A., *Deep Space Chronicle: A Chronology of Deep Space and Planetary Probes, 1958-2000* (Washington, DC: National Aeronautics and Space Administration Special Publication-4524, 2002).

18 Harford, James, *Korolev: How One Man Masterminded the Soviet Drive to Beat America to the Moon* (New York: John Wiley and Sons, 1997).

19 Miles, Marvin, "Dust Storm Smears Mariner Photos of Mars." *Los Angeles Times*, 11 November 1971.

20 Siddiqi, Asif A., *Deep Space Chronicle: A Chronology of Deep Space and Planetary Probes, 1958-2000* (Washington, DC: National Aeronautics and Space Administration Special Publication-4524, 2002), p. 86.

21 Ibid, p. 88.

22 Ibid, p. 88.

23 Harford, James, *Korolev: How One Man Masterminded the Soviet Drive to Beat America to the Moon* (New York: John Wiley and Sons, 1997).

24 Pickering, William H., "Odyssey to the Rim of the Solar System." Prepared for Pasadena *Independent Star-News*, Harold Wheelock Collection 1061-9, 1 January 1972.

25 American Institution of Aeronautics and Astronautics, Institution of Electrical and Electronics Engineers, and Society of Aeronautics Materials and Process Engineers.

26 Pickering, William H., "Thirty Trillion bits: Mariner 9 Maps Mars." Speech to combined meeting of AIAA/IEEE/SAMPE, Salt Lake City, Harold Wheelock Collection 1061-34, 24 January 1972.

27 Pickering, William H., "State of the Laboratory Message." Jet Propulsion Laboratory, Harold Wheelock Collection 1061-26, 16 March 1973.

28 Murray, Bruce, *Journey Into Space: The First Thirty Years of Space Exploration* (New York: W. W. Norton, 1989).

29 Mudgway, Douglas J., *Uplink-Downlink: A History of the Deep Space Network* (Washington, DC: National Aeronautics and Space Administration Special Publication-4227, 2001), p. 48.

30 Ibid.

31 In later years, the gravity-assist technique played a key part in several major deep space missions. Ulysses used the gravity fields of Jupiter to orbit the poles of the Sun; Voyager used the gravity of Jupiter, Saturn, Uranus, and Neptune to eventually reach the "rim of the solar system"; and the huge Galileo spacecraft depended on two gravity assists from Earth and one from Venus to reach Jupiter in 1995. Pickering observed all of this amazing technology with great interest and satisfaction, but he was not part of it.

32 Pickering, William H., "New challenges for the Engineer." Jet Propulsion Laboratory, Wheelock Collection 1061-83, January 1974.

33 Pickering, William H., "Improving the Environment." Jet Propulsion Laboratory, Wheelock Collection 1061-25, January 1974.

34 Pickering, William H., "American-Armenian International College." Jet Propulsion Laboratory, Wheelock Collection 1061-92, May 1974.

35 Pickering, William H., "Digitizing the Social Conscience." Jet Propulsion Laboratory, Wheelock Collection 1061-80, May 1974.

36 Pickering, William H., "Toward a New Society." Jet Propulsion Laboratory, Wheelock Collection 1061-1, May 1974.

37 Pickering, William H., "Inheriting the Future." Jet Propulsion Laboratory, Wheelock Collection 1061-82, June 1974.

38 Pickering, William H., "The Next One Hundred Years." Jet Propulsion Laboratory, Wheelock Collection 1061-57/58, November 1974.

39 Pickering, William H., "Proposal for an International Solar System Decade." Jet Propulsion Laboratory, Folders 127, 127, June 1974, Office of the Director Collection 1959-1982, JPL 142 Archives and Records Center, Jet Propulsion Laboratory, Pasadena, California, 2004; see also *Astronautics and Aeronautics*, vol. 12, September 1974, pp. 22–23.

40 Pickering, William H., "International Solar System Decade." William H. Pickering Collection 1958–1976, JPL 214, p. 13. Archives and Records Center, Jet Propulsion Laboratory, Pasadena, California, 2004.

41 In retrospect, Pickering did see most of the essential elements of his ISSD proposal implemented when NASA's space program regained momentum in later years and expanded to include many international cooperative space programs to study the Sun, most notably Helios and Ulysses.

42 Box 12, William H. Pickering Office File Collection 1955-1976, JPL 186, Archives and Records Center, Jet Propulsion Laboratory, Pasadena, California.

43 Pickering, William H., "Homo Sapiens: One of a Kind?" Jet Propulsion Laboratory, Wheelock Collection 1061-56, June 1975.

44 Pickering, W. H., Interview with the author. Pasadena, California, July 2003.

45 Folders 671, 672, Office of the Director Collection, 1959-1982. Pasadena, California: Archives and Records Center, Jet Propulsion Laboratory, JPL 142, 2004.

46 Ibid.

47 See "Lab-Oratory 1976-4," R. C. House Ed., for a report of the Pickering farewell events at both locations, Jet Propulsion Laboratory, 1976.

48 The Surveyor 3 camera, a very rare artifact indeed, eventually found its way to the Air and Space Museum in Washington, DC, where it remains on display in the section on "Exploring the Planets." See online at *www.nasm.si.edu/collections/space/space.html*, accessed 7 July 2007.

49 Koppes, Clayton R., *JPL and the American Space Program: A History of the Jet Propulsion Laboratory* (New Haven: Yale University Press, 1982), p. 241.

50 "Pickering: So Much, So Well." Part B, p. 6, *Los Angeles Times*, 1 April 1976.

51 Pickering, W. H., Interview with the author, Part 7A. Pasadena, California, August 2003.

Chapter 9

An Active Retirement

Saudi Arabia

By 1975, the College of Petroleum and Minerals near Dhahran in the Kingdom of Saudi Arabia had grown in size and status to the point where, by a royal decree, it was raised to the status of a university. The student body had expanded from about 60 to over 2,000 full time students at both the under graduate and the graduate level. Saudi and foreign students from over 40 countries were represented on the university's large desert-based academic campus. The full-time academic faculty numbered about 400, about half of whom were Saudi with the balance being foreign source academics under contract to the university.

The university looked to what it called a "Consortium of American Universities" for advice and guidance on a broad range of topics relevant to the establishment and operation of a first rank university. Representatives from Caltech, Massachusetts Institute of Technology, Princeton University, Stanford University, and the University of Michigan, among others, were part of the consortium, which met twice a year in Saudi Arabia or in the U.S.[1]

Plans for the following decade included a research institute devoted to applied research and development projects in five areas of specific interest to the Kingdom: namely water, oil, minerals, energy, and standards of measurement.[2]

Therefore, it was hardly surprising that in 1975, when the newly chartered University of Petroleum and Minerals (UPM) began looking for a director for its new Research Institute, it would send its Senior Advisor to the Rector[3] to talk with William Pickering, retiring director of the Jet Propulsion Laboratory (JPL).

Dr. Robert King Hall was "the guy from Saudi Arabia" that "turned up" at JPL, much to Pickering's surprise, about the time that he had concluded that a return to academic life at Caltech after retirement would not be in his best interests. King Hall's offer of a position as the first director of the new Research Institute in Saudi Arabia suddenly offered an intriguing alternative.

197

For the next several months, while the retirement functions swirled about him, Pickering considered the King Hall proposal. Ultimately, he agreed to accept a contract for two years, beginning in September 1976 when the UPM academic year began, right after the extremely hot months of midsummer were past. Since his function at this point was to be largely planning in nature, he did not see the need to reside permanently in Saudi Arabia. He would do his planning in Pasadena.

Pickering made the first move in his post-JPL career with a brief two week visit to the Research Institute right after the new director took office at JPL in April 1976. Impressed with the university and its infrastructure and, incidentally, intrigued by the cultural overtones that infused every aspect of Arabian academic life, Pickering returned to Pasadena to address the task of generating a long-term development plan for the institute. For this purpose he engaged several colleagues from Caltech and JPL on a part-time basis to help him plan a curriculum and formulate an operating mode for the new institute.

Pickering believed that the Research Institute should play a role for Saudi Arabia similar to that of the great American institutes of technology and his general plan, produced toward the end of 1976, reflected those concepts.[4]

The Research Institute would perform several very important research functions for the Kingdom. For the first time, engineering problems arising within the Kingdom could be investigated at home by a competent research team rather than by calling on experts from abroad. The Research Institute could initiate research activities leading to practical applications of particular value to the Kingdom and the unique institutional and technical resources of the Research Institute would enable it to perform the underlying research work that would lead to new industries in the Kingdom.

Most of the work would be done within the Research Institute, although some would be contracted out. Pickering emphasized the need to retain people with experience in applied research, principally from the U.S., as a resource for the Research Institute in its early years. He suggested criteria by which the institute could select engineering problems for study and criteria by which the approach to selected projects could be evaluated.

Finally, he suggested a management structure that was strongly representative of the flight project structure that he had advocated for so many years at JPL.

Pickering wrote, "Since much of the institute work will be oriented toward projects involving a number of technical disciplines, the staff will frequently be organized into project teams. These teams will be structured into a hierarchy of responsibilities, and their members will be required to work within the structure. Members must understand how to work as members of a team rather than as individual researchers as is usually true in university faculty research."

For the next couple of years Pickering divided his time between the Research Institute in Dhahran and his office in Pasadena, living on the university campus when in Saudi Arabia and living at home during his return visits to California.

It was a lifestyle that he thoroughly enjoyed and he took a great interest in the history, religion, and culture of the Arabic world in which he was now immersed. On one visit he took Muriel with him and together they shared a new and enriching experience of living with other expatriates and faculty in the "Compound" on the campus.

Between commutes to Dhahran and Pasadena he found time, and reason, to deliver speeches and lectures—on space in Saudi Arabia[5] and on "Islam and Oil" in California[6]—and to attend a conference in Lyons, France, on "Space and Civilization."[7]

When his contract with UPM expired in August 1978, Pickering returned to Pasadena and his protégé at the institute, a young Saudi scientist named Abdallah Dabbagh, took over as director of the Research Institute. By then, the institute was developing along the lines that Pickering had proposed in his original planning document. It was now up to the Saudi's to make it work and he had great confidence that Director Dabbagh had the potential to do that. From this point forward he would become a member of the university's Advisory Consortium and give the benefit of his advice once or twice each year as required.

In subsequent years, Pickering returned to Saudi Arabia on several occasions as a member of the Consortium.[8]

Over time, the Research Institute evolved into a vibrant state-of-the-art institution contributing its expertise in a wide spectrum of applied research to problems of prime interest to the Kingdom of Saudi Arabia. Furthermore, all of this had been accomplished within the original organizational structure.[9] William Pickering could justifiably take some considerable satisfaction in that.

Toward the end of his contract, William Pickering presented the university with a proposal to continue his association with the Research Institute as an independent technical consultant rather than as an employee of the university. To that end, he planned to set up a small, high tech consulting firm in Pasadena that would "provide manpower, expert consulting, technical studies, and supporting services to the Research Institute." In providing such support, the firm would, from time to time, draw upon experts readily available to Pickering through his close working relationships with numerous laboratories and educational institutions, most notably JPL and Caltech. The firm would be known as the Pickering Research Corporation (PRC) and would be based in Pasadena.[10]

With this business model in mind and high ambitions for future growth in a broad range of advanced technical areas, Pickering assembled a small group of professional colleagues from JPL and Caltech and with a very modest amount of working capital formed the Pickering Research Corporation: "a nonprofit institution to provide research and development support to the Kingdom of Saudi Arabia," in Pasadena in 1978.[11]

The precise details of the services to be supplied were to be negotiated in the form of a three year renewable contract between the PRC and the institute.

Research Institute of the University of Petroleum and Minerals, Dhahran, Saudi Arabia (Photo: Pickering Collection, Pasadena Museum of History, Pasadena, California).

Pickering's powerful influence in American technology, his extensive network of acquaintances in high office, and the appeal of his cause—build up of American technological influence in Saudi Arabia—assured him of ready access to expertise in a wide field of technology to implement his projects for the institute.

Plans and specifications for a satellite receiving station, an image processing center, and even an oceanographic institute were prepared.[12] Pickering did not limit the interests of the PRC to the Saudi contract, however, but sought other clients in Saudi Arabia and elsewhere. He formed agreements with several well-established high tech specialist firms in U.S. and made efforts to interest the Saudi government in weather modification experiments and environmental monitoring programs in which PRC would act as a facilitator between the Saudi users and the American suppliers.[13]

However, despite his good relations with Dabbagh and a prodigious amount of promotional work for each of these proposals, none of them came to fruition and by mid-1979 when his consulting contract with the Research Institute expired his connection with UPM reverted to his obligations as a member of the Consortium Advisory Board.[14]

Technology Transfer

As his involvement with the Saudi Research Institute wound down, Pickering turned his attention to finding other opportunities for his small corporation. Despite his disappointing experiences in "technology transfer" during his final years at JPL, Pickering retained a deep concern for the viability of the basic concept. As a result, space technology transfer soon became the basis for two new ventures in applied technology—one of them in nuclear power generation and the other in image processing—and both of them ultimately successful.

The Electric Power Research Institute (EPRI) was established in 1973 as an independent, nonprofit organization designed to manage a broad public-private collaborative research program on behalf of the electric utility industry, the industry's customers, and society at large.[15] Pickering and its founder Professor Chauncey Starr were old friends who "knew the ropes" and held a deep respect for each other's experience and accomplishments.

Pickering perceived that the nation's nuclear power generation industry had much in common with U.S. space and missile programs in the areas of reliability and safety and that the power generation industry might well benefit from knowledge of the "lessons learned" and the systems engineering approach that had evolved over the past 20 years in those two major national programs. His argument elicited a receptive response from EPRI and a proposal and contract soon followed. The deliverable was to be a report "documenting the evolutionary development of reliability and safety practices in unmanned and manned space projects and military intercontinental missile projects, over the last two decades." It would describe successes, problems, failures, and how they were corrected to increase the reliability of advanced systems of ever-increasing complexity. It would present the lessons learned in a way that would make it useful to other organizations—such as those engaged in nuclear power generation where high reliability in extremely complex systems was of the utmost importance.

Less than six months after his initial agreements with Chauncey Starr, the PRC completed its contract with EPRI. A few months later the Nuclear Safety Analysis Center published Pickering's report in a formal document for distribution throughout the industry.[16]

Riding on the success of his first contract with EPRI, Pickering followed up with a second proposal to the Nuclear Systems and Materials department of EPRI. Pickering proposed to describe the failure reporting systems then being used by JPL in the Voyager project and by the military in a major space project. These proven techniques, he believed, would have direct application to the nuclear power industry. Again EPRI accepted his proposal and, with help from a slightly different mix of consultants, PRC delivered "on time and in budget."

These were small contracts to be sure, but they served to sustain Pickering's interest in the viability of his tiny corporation, encourage him in the potential for "technology transfer," capitalize on his prestigious name, and provide an impressive front from which he could lever the organization into much more substantial contracts.

Searching for another niche in the field of technology transfer, Pickering turned his attention to computer-aided image processing.

As a consequence of its long experience with the challenges of data processing for its deep space projects, particularly Mariner and Viking, JPL had by then built up a substantial reservoir of unique expertise in the rapidly-advancing new technology of computer-aided image data processing—perhaps the best and most advanced in world at that time—and it was this expertise that it employed to support the Viking Mars mission.[17] Pickering, of course, was well aware of this unique source of expertise and knew exactly where to tap into it—in JPL's Mission Control and Computing Center—for the "technology transfer" he required for his next project.

Pickering planned to apply JPL's image processing technology to the manipulation of Landsat images to produce marketable products of interest to U.S. or foreign agencies engaged in the interpretation of surface imaging data for their particular fields of interest. Such areas of interest included demographics, geodetic or geographic features, vegetation, population, water prevalence, erosion, disaster effects, urbanization, surveying and cartography, agriculture, forestry, desert sand movement, etc., the list went on and on.

Landsat images were already readily available to public users as part of the government move toward commercialization of the Landsat image products.[18] However, to manipulate them and combine them with tabular and other graphic data to meet the specific requirements of individual users, powerful computers, advanced software programs, and peripheral image processing equipment were needed. PRC solved these and other problems by leasing the necessary JPL-developed software programs from the University of Georgia's Computer Software Management and Information Center (COSMIC) and renting time on suitable computers at Caltech's computer center. To manage the Landsat image data processing functions, he persuaded one of JPL's leading software experts to support PRC on a consulting basis as required. Pickering himself undertook to find customers for the esoteric services his corporation could now offer.

Initially, Pickering's Landsat image processing business involved relatively small contracts for U.S. customers who were interested in environmental and demographic problems. However, the returns from this work were not great and barely sustained the effort. From time to time he submitted a number of proposals to potential customers in Kenya, Saudi Arabia, Thailand, and Sri Lanka. In some cases, his proposals included a complete image processing center with training courses for personnel to manage, operate, and maintain it. None of these proposals, however, was successful.

Then, in March of 1980, as part of the gradually expanding cultural and scientific exchanges with the U.S., Pickering paid a visit to the Peoples Republic of China at the invitation of Dr. Chen Jie, Secretary General of the Chinese Society of Astronautics. He toured numerous academic institutions and manufacturing facilities associated with China's emerging space program and aeronautics industry and gave presentations on Mars, Jupiter, and spacecraft design. His visit proved to be of immense interest to the Chinese engineers and scientists. When he was shown a ground terminal for the Chinese experimental communications satellite, he elicited some discussion of Landsat data processing, a topic that held considerable interest for his Chinese hosts. Sensing some possible business opening in this area, Pickering left his hosts with a draft proposal to supply the Beijing Research Institute with an American-built image processing system.

Summing up his impressions later, Pickering thought that "the places we visited were very backward by our standards; they are short of good instruments, electronics, and computers, and have trouble in developing reliable systems and understanding trade-offs. They also appeared to be deficient in ground tracking and trajectory calculations."[19]

A few months later, Pickering submitted a full proposal from PRC to "design and install an image processing system for the Beijing Research Institute for remote sensing."[20] The price and delivery details were contingent upon obtaining an export license from the U.S. government.[21]

At first, the U.S. government authorities rejected his request for an export license on the grounds that some of the computing equipment and the JPL-developed software contained advanced data processing capabilities that conflicted with government policies regarding the export of U.S. technology to foreign countries. Pickering appealed to his Congressman for reconsideration. In the outcome, Pickering was forced to downgrade the hardware and software to the bare minimum required to manipulate Landsat images and resubmit his application.

Eventually, the long-delayed export license was approved and all of the hardware and software reached the end user in Beijing. In mid-1984 Pickering returned to China for a formal "acceptance and hand-over" ceremony to transfer the image processing system from its supplier, PRC, to its ultimate owner, Beijing Research Institute.

Through all of this frustration, Pickering continued to seek further image processing related business opportunities in China, including a proposal to supply a complete Landsat ground station to the Chinese government. Although none of these ventures was successful he, nevertheless, retained good relations with many top level executives in academic and government circles in that country and made reciprocal visits with them on several occasions in later years.

In the early 1980s, just as the outlook for a successful outcome for the contract with China was in serious doubt and, indeed, the future of PRC appeared doubtful, a colleague from JPL—a chemist by profession—drew Pickering's attention to an idea for an alternative fuel that involved the use of highly compressed pellets of wood waste-sawdust. At the time, the concept had already been patented but attempts to market the product had proved singularly unsuccessful.[22] Pickering picked up on the idea and in due course negotiated an arrangement with the owner of the patent to promote and market the idea through PRC.

Pickering soon began to focus his formidable promotional skills on developing a market for "Frajon pelletized fuels" which, at the time, were being produced very inefficiently, Pickering thought, by a plant in the St. Louis area. He sent out promotional material and made presentations, attended trade-shows, and cultivated contacts in the environment and air quality control industries and government agencies to draw attention to the highly efficient, clean burning, nonpolluting, combustible qualities of "pelletized fuel," but the response was not encouraging. [23]

Then began a remarkable chain of events that, many years later, William Pickering recalled in his inimitable, low-key style. "So about this time I got a call from a man in Idaho who said, 'I've got this pellet manufacturing plant [in Idaho] and it is about to go bankrupt, but it's a good idea and I need some money and I understand you are interested in pellets.' He hoped I would take an interest in his plant and bail him out."[24] But Pickering did not have the capital to intervene and, although he remained interested, the bank eventually foreclosed on the facility.

Acting on his intuition that pelletized fuel held great promise as an alternative source of energy, Pickering assembled a small group of colleagues of like mind into yet another small investment company with the idea of buying the foreclosed manufacturing plant from the bank at a favorable price and turning it into a profitable enterprise. The new company would refurbish the Idaho plant to improve its efficiency and operation and begin producing pellet fuel under the "Frajon" system. Pickering believed that by applying his redoubtable resources in technology to the manufacturing process, he could develop a dominant position in the then-faltering pellet fuel industry. Fortunately, he was able to negotiate an innovative financing arrangement that covered one year of refurbishment work and gave the Pickering group an option to eventually buy the plant under a longer term mortgage arrangement.

Pickering was also fortunate in finding a former JPL colleague with the appropriate expertise, experience, and availability to carry out the refurbishment work at the Idaho plant. Just over a year later, the work was done and the new plant at Sandpoint, Idaho, a few miles from the Canadian border, was in production. In 1983, Pickering renamed the new corporation "Lignetics" after the Latin word for wood and, with himself as Chairman of the Board and a small board of directors comprised of the dozen or so partners, opened Lignetics, Inc. for business.

As Chief Executive Officer, Pickering's devoted his full attention to the day-to-day management of the Idaho plant, promotion and distribution of the Lignetics product, and oversight of the tenuous financial stability of the fledgling corporation. For a man who, a few years earlier, bore ultimate responsibility for the disposition of an assured annual budget numbering in the many hundreds of millions of dollars and a professional staff of thousands, the task must have seemed very ordinary indeed. Yet, he found a personal challenge in it and was happy to employ his formidable intellectual resources toward making it succeed.

High Honors

JPL's spectacular retirement functions in March of 1976 did not represent the end of public recognition for William Pickering's pioneering career in deep space. For the next 25 years, public and professional institutions, national and international, accorded him their highest honors.

In November 1975, Queen Elizabeth II of England had conferred upon William Hayward Pickering the distinction of Honorary Knight Commander of the Most Excellent Order of the British Empire for his "Services to Science." The "Honorary" designation signified his formal, non-British citizenship and entitled him to wear the insignia and append the nominals K.B.E. to his name. Henceforth, his formal address would become "Dr. William H. Pickering, K.B.E." The new title assured him of a permanent place in the ranks of other great New Zealand achievers.

In early June 1976, he and Muriel made a short visit to New Zealand for the investiture, held in this case by the Queen's representative in New Zealand, Governor-General Sir Denis Blundell at Government House in the capital city of Wellington. The short ceremony was followed by an official luncheon at Government House.

Before returning to California, Pickering filled a long-standing commitment to address the 50th anniversary convention of the New Zealand Association of Radio Transmitters—the country's leading amateur radio organization. As New Zealand's most famous living "Ham,"[26] he represented a direct link to the early days of radio at Wellington College. His address was appropriately titled "From Galena to Silicon," a clever reference to the technological advance from early radio based on galena crystal detectors to modern silicon-based transistor radio.[27]

Later that year, in company with his friend Wernher von Braun and highly respected colleague Frederick Terman, William Pickering paid another visit to the White House in Washington, DC. On this occasion, 18 October 1976, he was to receive the National Medal of Science from President Gerald Ford.[28]

This was the nation's highest honor for engineering excellence and its recipients were giants of American technology who would be worthy bearers of the honor that had first been bestowed upon Pickering's mentor, Theodore von Kármán. The National Science Board citation read: "For his leadership

Dr. William H. Pickering, K.B.E., with Sir Denis Blundell, Governor General of New Zealand, following the investiture ceremony in Wellington, New Zealand, 2 June 1976 (Photo: Courtesy of the Pickering Family Trust).

of the exploration of the planets of the solar system and his personal contributions to the theory and practice of soft planetary landings and the collection of data from deep space." Conferred by the nation's highest executive and endorsed by his peers, the National Medal of Science was perhaps the award that he valued most for it represented, unequivocally, excellence in his selected field of endeavor.

Pickering's propensity for international travel showed no sign of diminishing after he left JPL. For the first few years he made numerous visits to Saudi Arabia as part of his relationship with the University of Petroleum and Minerals in Dhahran. On occasion, he extended these trips to include visits to India, Sri Lanka, and Thailand to engage in discussions, generally accompanied by lectures, on the U.S. space program and related topics at universities or technical institutes. He gave papers and lectures in France and Germany for similar purposes.

The Australia and New Zealand Association for the Advancement of Science (ANZAAS), New Zealand Amateur Radio Transmitters (NZART), New Zealand Institute of Management, Center for Advanced Engineering (CAE), Institution of Professional Engineers New Zealand, Royal Society of New Zealand, and the Canterbury School of Engineering were at one time or

William H. Pickering receives the 1975 National Medal of Science for Engineering from President Ford at the White House, Washington, DC, 17 October 1976 (Photo: Pickering Collection, Pasadena Museum of History, Pasadena, California).

President Ford chats with Dr. William Pickering and his wife Muriel at the White House reception following the presentation of the National Medal of Science, 17 October 1976 (Photo: Pickering Collection, Pasadena Museum of History, Pasadena, California).

another honored with personal appearances and presentations by William Pickering. Affording him an opportunity to visit family and friends, institutionally sponsored visits to New Zealand were always an acceptable obligation in those years.

In his retirement years, Pickering's travel within the U.S. was less frequent and far less onerous than it had been in JPL years. Requests for appearances and speeches diminished as NASA's achievements in space no longer held the public interest they once did. In any case there were new personalities at NASA and JPL to talk about the remarkable Voyager encounters with Jupiter and Saturn, Uranus and Neptune, and the new projects to explore the solar system. Gradually Pickering's speeches began to address topics in the field of alternative energy. "Conservation and New Energy Alternatives," "Densified Wood as Domestic Fuel," "A New Energy Resource–Wood," and "Pellet Economics" were examples of this new trend in his thinking.[29]

Throughout his professional life he had maintained a strong connection to the American Institute of Aeronautics and Astronautics (AIAA). Together with cofounder L. Eugene Root, he had brought together the American Rocket

Society and the Institute of Aerospace Science to establish the AIAA in 1963 and he had acted as its first president. He had exercised his considerable prestige and wisdom to formulate the guiding principles of governance that enabled the AIAA to grow to become the world's leading organization of aeronautics and astronautics professionals with an international reputation and active alliances in many European and Soviet countries. His personal contribution to its programs and his powerful influence on the organization were a matter of record.[30]

In 1986 the AIAA recognized William Pickering's contribution to the advancement of aerospace technology with a celebratory dinner at the Caltech Athenaeum and the presentation of its 1986 Aerospace Pioneer award. Among many laudatory messages he received on that occasion were letters from U.S. President Ronald Reagan, California Governor George Deukmejian, National Academy President Frank Press, and many others representing NASA, universities, distinguished scientists, and the country's major aerospace industries.

Pickering's personal satisfaction with the recognition he received as a consequence of his success in public life in these years was tempered by a grievous setback to his personal life in 1992 when, after a prolonged illness, Muriel died of congestive heart failure.

Throughout all of the 59 years that they had been married, William Pickering looked to his wife Muriel for comfort and encouragement when his professional life was difficult and for recognition and approval when it was successful. She had traveled the world with him and shared his introductions to princes and presidents, prominent politicians, distinguished scientists, and dull people aspiring to public recognition by the act of meeting those who had already achieved it. To her husband's public charisma she added her own natural charm, equanimity, and graciousness.

William Pickering bore the loss of his beloved wife with a private grief that those who knew him personally were not party to, for in personal matters he was an intensely private man. He was also a sensitive man, and gentle, and his grief, if not visible, was assuredly palpable. The La Cañada-Flintridge community paid handsome tribute to Muriel's generous contributions to its welfare, law enforcement, beautification, and city improvement activities. She was sorely missed in many ways.

In the months that followed, his daughter Beth helped her father to deal with the loss that they both felt deeply and to adapt to his changed personal circumstances, while Pickering immersed himself in the demands of his new business venture.

Beyond his business interests, Pickering's prominence in the field of aerospace engineering continued to expand as further high honors were bestowed upon him—on this occasion from a European source.

Established in 1992 by European philanthropist Albina du Boisrouvray in memory of her son, the François-Xavier Bagnoud Aerospace Prize was largest international prize in its field. It was to be administered by the university of Michigan's department of Aerospace Engineering at Ann Arbor, Michigan, where her son had taken his degree in aerospace technology.[31]

The following year the prize selection committee, comprising of aerospace experts from around the world, nominated William Hayward Pickering to be the first recipient of the valuable François-Xavier Bagnoud Aerospace Prize. At his nomination, Pickering commented, "I am honored and delighted to be chosen as the first recipient of the François-Xavier Bagnoud Aerospace Prize and I am a bit overwhelmed by the honorarium [$250,000]. Thank you very much indeed. Being the first recipient is particularly significant to me when I look at the truly international membership of the selection committee."

In a background paper on William Hayward Pickering, Thomas E. Everhart, then president of Caltech wrote, "More than any other individual, Bill Pickering was responsible for America's successes in exploring the planets—an endeavor that demanded vision, courage, dedication, expertise, and the ability to inspire two generations of scientists and engineers at the Jet Propulsion Laboratory."[32]

Pickering's presentation dinner address "Some Reflections on Space Research: The Challenges and the Triumphs," traced the evo-lution of the U.S. space pro-gram from the initial events leading up to Explorer 1 and the early Moon and Venus and Mars missions, through the Voyager missions to Jupiter, Saturn, Uranus, and Neptune. It was, in essence, a retrospective of his own life told on this occasion from the vantage point of history rather than, as in the past, that of a real time participant. "The space program of the 1960s and 1970s was a major weapon in the Cold War," he said. But public support for space was not what it was during the Apollo period.

William Pickering with his daughter Beth Pickering Mezitt, Los Angeles, April 1992 (Photo: Courtesy of the Pickering Family Trust).

"Thirty years later, a fickle public finds space boring. With a recession and an enormous government debt, the cost of space research looks like a controllable area of the national budget that can be sharply reduced. Space flights today seem to be noteworthy only when disaster strikes and the results of space missions seem to be of interest only to a small group of scientists, and are considered by the public to be of no practical value," he said. But we should be looking ahead. He concluded:

> We should regard the ability to explore space and to travel in space as another step in the evolution of mankind, upwards from the cave. My generation took the first steps in space—your generation is the first to have grown up with the knowledge that man has stepped off Earth into the cosmos. You cannot go back—you must move forward. As the Russian rocket pioneer K. E. Tsiolkovsky said almost 100 years ago, 'Earth is the cradle of mankind, but one cannot live in the cradle forever.'

On this occasion, his pleasure was enhanced by the presence of his daughter Beth and her husband and the company of a longtime family friend from La Cañada by the name of Inez Chapman. Inez, it turned out, was destined to play a major role in William Pickering's later life, but for now other matters demanded his immediate attention, foremost among them an invitation to visit the Emperor of Japan.

William Pickering marked his 84th birthday in December 1994 with a few friends and dinner at the beautiful Athenaeum on the Caltech campus in downtown Pasadena. As a Professor Emeritus he enjoyed the deference accorded his visits to the Athenaeum and the ambience of the grand dining hall spoke to him eloquently of the great men of science, many of them his personal friends, who had passed through it in times past.

That evening, Pickering's birthday party also celebrated his selection as a recipient of the 1994 Japan Prize for Aerospace Technologies. The prize carried with it an award of 50 million yen.[33] It was certainly a sufficient cause for celebration.

He was in good company. The other recipient of a Japan Prize that year would be a Dr. Arvid Carlsson a distinguished pharmacologist from the University of Gothenburg, Sweden. Established by the Science and Technology Foundation of Japan in 1985, the Japan Prize demonstrated Japan's commitment to science and technology by honoring those in the international community whose work had contributed to the peace and prosperity of humankind.

For Inez Chapman, now his fiancé and companion on the visit to Japan, the social events in the elaborate program presented, understandably, new challenges. To prepare herself for the task ahead, Inez took time to inform herself about the Japanese royal family and its background and the customs, culture, and recent history of Japan. For his part, Pickering prepared the "commemorative" lecture titled "Space Technology: The New Challenge."

The formalities of "Japan Prize Week" included a joint press conference, receptions at the American and Swedish Embassies, a visit with the prime minister of Japan, the Japan Prize presentation ceremony in the presence of His Majesty the Emperor of Japan, a formal banquet attended by their Majesties the Emperor and the Empress, and commemorative lectures and discussions with an international panel of scientists and technologists.[34]

In his commemorative address, Pickering traced the history of the U.S. space program from its inception, shortly after the debut of Sputnik in 1957, to the present time, reminding the audience that the initial exploration of space was accomplished by both the U.S. and the Soviet Union. Later, other countries, including Japan became involved with the U.S. programs. As scientific questions became more detailed and the missions became more complex, greater opportunities for international cooperation arose.

Pickering expressed his pleasure at the extent to which Japan was now engaged in cooperative space ventures with JPL and expressed the hope that "such cooperation might soon include many other space faring nations as well." In conclusion, he said, "There are worlds that have waited for us for billions of years. The end of the waiting period has begun. We are the first generation to open the doors for the generations to follow."[35]

In addition to the very substantial prize money, William Pickering received an exquisite gold medallion that symbolized the Sun, national emblem of Japan, and the perfection of the circle. "For inspirational leadership in unmanned lunar and planetary exploration, and for pioneering achievements in the development of spacecraft and deep space communications": a fitting tribute from a nation that held such works in high esteem.

For Inez the visit to Japan and the events associated with it had been the experience of a lifetime. It was a difficult act to follow, but follow it she did, by accepting Bill's offer of marriage and celebrating it a few months later with a splendid wedding in the La Cañada Presbyterian Church on 27 July 1994. Bill Pickering's "cup of happiness" was full indeed.

The Emperor of Japan Akihito greets William Pickering and Inez Chapman on the occasion of the Japan Prize presentation banquet, Tokyo, Japan, April 1994 (Photo: Courtesy of the Pickering Family Trust).

The new couple settled down in La Cañada-Flintridge, living in the smaller of the two houses they then owned, and set up a combined household that satisfied the needs of both. For Inez that meant maintaining her piano teaching classes and her community and social activities; for Bill it meant sustaining his business interests in Lignetics and maintaining his numerous obligations to professional institutions such as the Academy of Science, AIAA, and IEEE. It all worked out very well indeed.

The new Lignetics plant in Glenville, West Virginia, had come on line that year. A 1997 news release announced its opening: "Dr. William Pickering, Chairman, Lignetics Inc., informed stockholders in November that the company's new West Virginia pellet fuel manufacturing plant is now in production. This makes Lignetics the largest producer of wood pellet fuel in the United States with combined [three-plant] production of capacity of 200,000 tons per year. This is more than three times the amount of the industry's second largest competitor."[36]

Now, with three pellet plants in production, the Lignetics organization required Pickering's strong direction to make the enterprise profitable. No longer involved with interplanetary space missions and the politics of government/university relations, Pickering's attention was fully engaged with the intricacies of financial statements, production goals, overhead costs and distribution, seasonal inventory, and the price of wet sawdust.[37]

Space Revisited

Despite his business interests, however, William Pickering's interest in deep space exploration had not entirely waned, nor had his name and charisma lost its ability to attract an enthusiastic audience.

When JPL and Caltech celebrated 40 years of space exploration in January 1998 with displays from the past and present and a public meeting at Caltech, the featured speakers were William Pickering and James Van Allen,[38] surviving members of the original Explorer 1 group, and Edward Stone current director of JPL. Pickering spoke of the momentous events surrounding the Explorer 1 event of so long ago, while Stone spoke of the present and the outlook for the future. The three speakers epitomized the amazing progress that spanned the first 40 years of JPL's venture into space exploration. Many of those present were hearing the Explorer 1 story from the legendary participants themselves for the first and, probably, the last time.

Pickering recreated the dramatic events of the launch of Explorer 1—the first detection of the signal from the satellite that signified America's first satellite was in orbit and the tumultuous media reaction that followed the realization that Americans had, at least for the moment, caught up with the Russians in the race for space.[39] In the rush of public excitement it was Explorer 1 versus

Sputnik 1—the U.S. versus the USSR. And there was more, as Pickering was quick to point out. We had science, real and not pseudo, from Van Allen's unique radiation instrument.[40]

The Explorer 1 satellite, conceived and built in haste at JPL, carried with it the hopes and expectations of Cold War America, a country reeling from the shock of the Sputnik demonstration of Soviet superiority. Looking back from 40 years, Pickering could, with some justification, have pointed to the influence of science and technology as a powerful factor for the betterment of humankind, a claim that had been the focus of much of his public discourse in the early years.

"NASA Failures Blamed on Policies," proclaimed the *Los Angeles Times* in March 2000.[41] Suggesting that "NASA took undue risks by working on major space missions too quickly and too cheaply," and this article, like many on that date, referred to the failures, just about one month apart, of two spectacular missions to Mars: the Mars Climate Orbiter and the Mars Polar Lander both of which were the responsibility of JPL. Reminiscent of the early Ranger missions to the Moon, the mishaps had triggered a public outcry critical of NASA and JPL and prompted massive investigations by both organizations in addition to giving rise to a congressional enquiry.

From his modest Lignetics office just across the Arroyo Seco, Pickering empathized with the stress and anguish that the situation brought to the current occupant of his former office at JPL. Forty years earlier he had been in that "hot seat," but now he was on the outside looking in and now, as then, he moved quickly to voice his opinion on the nature of the problem. More than 20 years absence from JPL had not diminished his legacy or dulled his influence within the JPL-Caltech culture. Soliciting opinions from a number of his former senior engineering managers, Pickering quickly produced a position paper titled "Comments on Recent JPL Failures."[42] Referring to the Ranger program and the JPL "culture" that it engendered, Pickering wrote:

> The engineers and scientists of that day were not smarter than they are today . . . but they did some things differently. JPLers were confident, proud of their skills—some even said arrogant. They were cautious, they were meticulous. NASA had difficulty understanding the JPL staffers who were so different from their civil service counterparts.

Pickering described the "loose hierarchical system" that encouraged close communications within the JPL organization and emphasized the importance of "system design and attention to engineering detail" in JPL's success. Since the only real test of a complete (spacecraft) system occurs with the flight itself, "preflight testing must come as close as possible to the actual flight conditions," he wrote.

Pickering believed that to recover from the trauma of losing two Mars missions, JPL must first review its relationship with NASA to restore "both authority and responsibility for technical detail." Changes were also required within JPL. Perhaps, he surmised, the "matrix" type of organization that had worked so well in the past for a relatively small number of large projects was no longer suited to the current work programs that comprised a larger number of smaller projects. The organizational structure within JPL should be reviewed to find the answer to that question.

Sensitive to the potential value of ideas at all levels within JPL, Pickering asserted that "the internal organization should welcome technical discussions with the younger generation . . . who have sparked the technology boom," and he encouraged the younger generation to "take advantage of the hindsight that the history of NASA and JPL offered them."

Much of Pickering's thinking about NASA and JPL was all too familiar. To the "older generation" it seemed that the problems were the same, only the people had changed. Eventually, as it had done in the past, JPL righted itself and went on to greater glory, leaving behind yet another wrinkle in the rich fabric of Pickering's life experience.

Endnotes

1 Research Institute, ". . . And Thus Seek Knowledge." Dharan, Saudi Arabia: University of Petroleum and Minerals, January 1976.

2 Research Institute, "The Ultimate Resource." Dharan, Saudi Arabia: University of Petroluem and Minerals, November 1979.

3 The chief executive officer responsible for the implementation of policy and administration of the university.

4 Pickering, W.H., "The Research Institute." Undated, 1976. In the Pickering Collection, Pasadena Museum of History, Pasadena, California.

5 Pickering, W.H., "Engineering in Space." Society of Petroleum Engineers, Dharan, Saudi Arabia, 25 January 1977; "The Landsat Satellite." University of Petroleum and Minerals, Dhahran, Saudi Arabia, 20 June 1977. In the Pickering Collection, Pasadena Museum of History, Pasadena, California.

6 Pickering, W.H., "Islam." La Canada Universalist Church, La Canada, California, 19 April 1977; "Oil, Sand, and the Koran." Rotary Club, Pasadena, California, 20 June 1977. In the Pickering Collection, Pasadena Museum of History, Pasadena, California.

7 Pickering, W.H., "Exploring the Planets—The First Steps." IAF Conference on Space and Civilization, Lyons, France, 7 June 1978. In the Pickering Collection, Pasadena Museum of History, Pasadena, California.

8 Research Institute. ". . . And Thus Seek Knowledge." Dharan, Saudi Arabia: University of Petroleum and Minerals, January 1976.

9 Dabbagh, A. E. "Research Institute Progress Report." Dhahran, Saudi Arabia: King Fahd, University of Petroleum and Minerals, January 1989.

10 W. H. Pickering to Abdullah Dabbagh, Rector, Dharan, Saudi Arabia: University of Petroleum and Minerals, 23 July 1978. In the Pickering Collection, Pasadena Museum of History, Pasadena, California.

11 Pickering Research Corporation, "Proposal," 17 June 1978. In the Pickering Collection, Pasadena Museum of History, Pasadena, California.

12 Pickering W.H. "Planning documents prepared for the R.I, UPM, 1979: Image Processing Center, Oceanographic Institute, Satellite Receiving Station." 30 April 1979. In the Pickering Collection, Pasadena Museum of History, Pasadena, California.

13 W. H. Pickering to Abdullah Dabbagh, "Planning reports," 18 May 1979; "Invoice for consulting support and activities report," 1 June 1979. In the Pickering Collection, Pasadena Museum of History, Pasadena, California.

14 Pickering, William H., "Progress reports prepared for the R.I." Annual Report, 1 September 1978 to 31 August 1979; "Report for the Director," March 1980; "UPM Consortium Visitation Report," 6–18 March 1980. In the Pickering Collection, Pasadena Museum of History, Pasadena, California.

15 See about APRI online at *http://my.epri.com*, accessed 7 July 2007.

16 Pickering Research Corporation, "Space and Missile Reliability and Safety Programs." Nuclear Safety Analysis Center, Doc. NSAC 31, February 1981.

17 Each Viking spacecraft comprised an Orbiter and a Lander. The Orbiters were equipped to make high definition images of the Martian surface from orbit, while the Landers were to carry out their photographic missions from the surface. JPL would provide tracking and data acquisition services including the imaging processing functions.

18 Landsat 1 was launched in 1972 and, over the next two years, transmitted over 100,000 images covering 75 percent of Earth's land surface to the tracking and data receiving center located at the Goddard Space Flight Center in Greenbelt, Maryland. These data were made available to over 300 investigators in the U.S. and foreign countries.

19 Pickering, W. H., "Notes on Visit to China," 27 March to 7 April 1980. In the Pickering Collection, Pasadena Museum of History, Pasadena, California.

20 Pickering, W. H., "Pickering Research Corporation Proposal to Beijing Research Institute," 1 August 1980. In the Pickering Collection, Pasadena Museum of History, Pasadena, California.

21 Before embarking on this project, Pickering sought unofficial opinions from his connections in the Pentagon and elsewhere in Washington to the effect that such an application from PRC would be viewed with favor by the official authorities.

22 The concept of using pelletized sawdust as a fuel was not new—it had been around for some considerable time. However, the patented "Frajon" process eased the manufacturing process and substantially improved the combusting quality of the product by the use of a special additive that was the subject of the patent.

23 Pickering, W. H., "Demonstration of Frajon Fuel Pellets as a Direct Replacement for Oil in Oil-Fired Boilers." Unsolicited Proposal to the Deputy Director Engineering and Technology, Office of the Corps of Engineers, Department of the Army, 5 March 1981. In the Pickering Collection, Pasadena Museum of History, Pasadena, California.

24 Pickering, W. H., Interview with the author, August 2003.

25 As a U.S. citizen, Pickering was not entitled to use the title "Sir," neither in New Zealand nor in any other country. (Honours Secretariat, N.Z. Prime Minister's Department, April 2005.) Although William Pickering never abrogated the privilege, it nevertheless had become common public usage in New Zealand to refer to him as Sir William Pickering.

26 Amateur radio operator.

27 Dougherty, Ian, *Ham Shacks, Brass Pounders and Rag Chewers: A History of Amateur Radio in New Zealand* (New Zealand: New Zealand Association of Radio Transmitters, 1997), p. 227.

28 See online at *http://www.nsf.gov/nsb/awards/nms*, accessed 17 May 2007.

29 From the Pickering Collection, Pasadena Museum of History, Pasadena, California.

30 Copies of Pickering's extensive correspondence with the AIAA (and other professional institutions) are contained in Pickering, William H., Records 1970–1989. Pasadena, California: Archives and Records Center, Jet Propulsion Laboratory, JPL 3, 2004.

31 University of Michigan, Notes on "Albina Du Boisrouvray," 23 September 1993. In the Pickering Collection, Pasadena Museum of History, Pasadena, California.

32 Everhart, Thomas E., "Notes for the François-Xavier Bagnoud Aerospace Prize." Ann Arbor Michigan, 7 October 1993. In the Pickering Collection, Pasadena Museum of History, Pasadena, California.

33 Equivalent to about $500,000 U.S. at the time the award was made (1994).

34 Science and Technology Foundation of Japan, "1994 10th Anniversary, Japan Prize Presentation Ceremony; Background," 27 April 1994. In the Pickering Collection, Pasadena Museum of History, Pasadena, California.

35 Pickering, W. H., "Space Technology: The New Challenge." April 1997. In the Pickering Collection, Pasadena Museum of History, Pasadena, California.

36 Lignetics, Inc, "Lignetics Announces New Pellet Production Milestone." *News Release*, 24 November 1997. In the Pickering Collection, Pasadena Museum of History, Pasadena, California.

37 The Missouri plant did not meet the company's expectations and was closed the following year.

38 Wernher von Braun the third member of the original group credited with the success of Explorer 1 had passed away in 1977.

39 Pickering, W. H., "40th Anniversary of Explorer 1." January 1998. In the Pickering Collection, Pasadena Museum of History, Pasadena, California.

40 Ill health prevented James Van Allen from traveling to Pasadena but he sent a video recording in which he told of his discovery of the now-familiar bands of intense radiation that surrounded Earth and now bear his name. Van Allen, James A., "Text of statement by James A. Van Allen for a gathering at Caltech on January 31, 1998." January 1998. In the Pickering Collection, Pasadena Museum of History, Pasadena, California.

41 McFarling, Usha Lee, "NASA Failures Blamed on Policies," *Los Angeles Times*, p. A3, 14 March 2000.

42 Pickering, W. H., et al., "Comments on Recent JPL Mission Failures." 10 March 2000. In the Pickering Collection, Pasadena Museum of History, Pasadena, California.

Chapter 10

Full Circle

Full Circle (1994–2004)

At age 90, William Pickering was an imposing figure. Tall, slim, and of upright bearing, he walked purposefully, unaided by cane or walker. He dressed plainly but well, informally or otherwise, and always with good taste. His graying hair, sparse but not bald, was always neatly combed in place and his facial features, unadorned by beard or moustache, were clear, healthy-looking, and well shaped. Piercing blue eyes beneath a prominent forehead imparted an air of authority to his presence that was a defining characteristic of his persona. Thin lips and a small, tight mouth gave his speech a clipped incisiveness that complemented his deep resonant voice and residual "Kiwi" accent. His sight and hearing were good and his memory and power of recollection were awesome. He did not smoke but enjoyed whisky, beer, or wine in moderation as the social occasion demanded. In conversation, he often summoned a tight-lipped Mona Lisa-like smile but seldom laughed outright, tending to "chuckle" deep within his throat when amused.

He bore his many honors like the elder statesman of science that, by that time, he really was. He possessed a personal charisma that endeared him to audiences wherever he went and was equally at home in the company of presidents or princes; Congressmen or kings; and preeminent scientists or adoring school children. The media loved him for he had the knack of pitching his remarks to the level of understanding of his listeners and he was never at a loss to answer a question. "Let me reflect on that . . ." he invariably began. The media called him, with justification, the "gentle giant of science." Among his peers throughout the world he engendered the greatest respect and among those he led at JPL he generated a loyalty that never waned. He was a natural leader and knew no other way to conduct the business of his life.

The new century found William Pickering taking a much less active role in his business affairs. The unprofitable Missouri plant had since been closed and the two remaining plants were operating profitably. A younger man who had been with the company since its inception carried the title of Chief Executive Officer and now directed the company's regular business from new headquarters in Idaho. Pickering retained the title of Chairman of the Board of Directors and maintained a modest office in Altadena just across the Arroyo Seco from JPL. There he periodically convened meetings of his stockholders and dealt with a continuous stream of daily business associated with his remaining professional interests. Although the company was growing slowly, Pickering felt that it faced an uncertain financial future that might, in the not too distant future, necessitate its sale to satisfy shareholder expectations for a more timely return on their investments. But for now, he was content to let events take their course and look to others to deal with the situation as it evolved.

Although Pickering's association with professional societies in the field of aeronautics and astronautics grew tenuous in his later years, the institutions that he had served so well in his earlier years—founder in one case, president in most others—did not forget him and they sought to demonstrate their appreciation with the bestowal of their highest awards.[1]

"In recognition of a distinguished career that pioneered and shaped the exploration of our solar system and, for extraordinary contributions to engineering and science," read the citation for the year 2000 Guggenheim Medal awarded to William Pickering by the American Institute of Aeronautics and Astronautics at its Honors Night banquet in May 2001. The John F. Kennedy Astronautics Award for Public Service is given to an individual who "has made an outstanding contribution by promoting the nation's space programs for the exploration and utilization of outer space," stated the American Astronautics Society in their 2002 award of this distinction to William Pickering. Accompanied by his wife, Pickering attended the award banquets with obvious delight and accepted the accolades with graciousness and inevitably, with an appropriate speech of acceptance.

Less now a bully pulpit for the future of space exploration, institutional reform, and the deployment of space technology for the betterment of humankind, Pickering's occasional speeches turned back to review what had gone before and how he had been involved in the opening of the Space Age. The speeches were shorter and less technical and, for audience impact, relied largely on stunning slide images from the Viking and Voyager missions to Mars, Jupiter, and Saturn rather than a mind-bending reiteration of what were, at the time, astounding facts of the new technology of space exploration. An address to the World Space Congress[2] on early exploration of the solar system in October 2002, and an opening address to an international telemetering conference[3] in Las Vegas in October 2003 were instances of this trend and were in fact to be among the last of his public presentations.

Like Pickering himself, those who had been old enough to recognize the threat of the Soviets' Sputnik coup and experience the tensions of the Cold War era had aged, and indeed many had been replaced by a new generation of young technologists for whom space was a given—an established fact—a point of departure for even greater achievements in space exploration. The space pioneers, Pickering foremost among them, were immortalized in many places throughout the country where feats and facts and fiction related to America's achievement in space provided attractions for educators, tourists, and, occasionally, fanatics. Pickering's portrait, achievements, and biography occupied prominent positions in the exhibition areas of such places as the International Aerospace Hall of Fame at Alamogordo, New Mexico, site of the first V2 rocket tests in U.S. and the International Aerospace Hall of Fame in San Diego, California, a key center for American aerospace development from early aircraft to modern booster rockets.

Pickering fulfilled a long-standing obligation when he returned to New Zealand yet again in March 2002, to attend a rededication ceremony for the recently restored Gifford Observatory at Wellington College. He was to be the primary speaker and guest of honor. It was an obligation that he gladly accepted. "Uncle Charlie" Gifford had been the resident mathematics teacher at the College during Pickering's formative years of 1923 to 1927, when he was a pupil at the school, and he had played a major part in establishing young Pickering's general interest in mathematics and science—physics and astronomy, in particular. He forever carried a debt of gratitude to the memory of Charles Gifford, whose gift to the school of a 5-inch Zeiss refractor telescope had made the wonders of astronomy a reality for generations of students like William Pickering.

Local media reported widely on the presence of Dr. Pickering and his wife in Wellington and the Pickering's visit very quickly expanded into a frenzy of public appearances, talks, and lectures

Headmaster Roger Moses greets Dr. and Mrs. William H. Pickering at Wellington College, Wellington, New Zealand, March 2002 (Photo: Neely Collection, Wellington College Archives).

interspersed with a public lecture to a crowded Town Hall,[4] the dedication for the Gifford Observatory,[5] and an address to the plenary session of the Institution of Professional Engineers New Zealand (IPENZ).[6] Newspaper and television interviews vied for time with the city's distinguished visitor. Fuelled by the attraction of his reputation and enhanced by his royal nominal (K.B.E.), his popularity in New Zealand could hardly have been greater.[7]

The following year, in March, the final details of John Campbell's plan for the Rutherford-Pickering Memorial in Havelock fell into place, and a date for the dedication ceremony was set. Accompanied by his wife and daughter, Pickering returned to New Zealand once more, this time as the guest of honor for the celebrations in Havelock.

Return to Havelock

Late summer in the Marlborough Sounds can be an enchanting time. The days are long, clear, and sunny. Gentle winds rather than the howling gales that roar through Cook Strait at other times of the year keep the temperatures in a comfortable range. The steep hillsides that cradle the long arms of sparkling water connecting to the sea are greened by the occasional overnight rain. Around Havelock, a profusion of wild flowers are in bloom and dairy cattle browse the lush green pastures. The flood of summer tourists has subsided and the smell of saltwater hints at Havelock's historic connection to the sea. Saturday, 15 March 2003 was just such a day—picture postcard perfect.

Along the main road that passes directly through the tiny township, several of the few shops that comprised the business area of Havelock displayed hand-printed signs reading "Welcome home Sir William" and "Havelock honors Sir William Pickering" and from the town hall in the center of town flags hung listlessly in the bright morning sunshine.

Up at the modern Havelock Primary School,[8] a short walk beyond the town, school children began assembling early for this was to be a special event—Havelock's own, world-famous scientist was coming to town, to their school, to talk to them.

In a short speech, matched perfectly to his juvenile, but very well-informed audience, he told stories of his school days at the old Havelock School and spoke of the wonderful adventures in space that had been the main part of his life. He admired the splendid murals and displays of the solar system that the students had prepared for the occasion. He encouraged them to study hard and to keep learning, saying that "what you do at school is important for what you do after school." Afterwards he patiently answered questions, signed autographs, and planted a tree to his memory. And then it was time to go on to the next event.

Along the main street, the tooting of car horns and shouts of welcome signaled the arrival of the guest of honor. Riding in an antique Ford Roadster, William Pickering, accompanied by his wife Inez and daughter Beth, swept into town—the ladies waving graciously to the astonished bystanders, the guest of honor gallantly acknowledging their welcome with a nod and a smile.

Following speeches from officials representing civic and Maori interests, Pickering spoke of his pride at being honored, along with Lord Rutherford, by the splendid memorial. He said:

> This memorial will be seen by many who will wonder who these people were . . . and perhaps come to believe that if two boys from Havelock can become world figures, perhaps they too, should strive to develop their talents. . . . Each in our own way, we exemplify man's insatiable curiosity to learn about his universe. Rutherford and I happen to have been in the right place at the right time to be able to use the talents we were blessed with to contribute to the encyclopedia of knowledge that describes the universe.

Saying that we now live in an era of instant communications that allow ideas and leadership to develop in even the smallest parts of a community, Pickering asked, "Who knows which of you will make a singular contribution to mankind and which of you will have a message for the world?"[9]

At the appropriate time and, somewhat jostled by the press of curious spectators, photographers, TV crews, and news reporters, Pickering cut a ribbon to unveil the memorial panels and, to a burst of polite applause, the Rutherford-Pickering Memorial was declared open. A group of school children sang "God Defend New Zealand," New Zealand's national anthem, and the ceremonies closed with a magnificent New Zealand country luncheon in the nearby town hall. William Pickering's boyhood friend Ulrich Williams, now also 94 years of age, joined him at the luncheon and the two old friends enjoyed reminiscing about times long past, while both savored the wonderful food that, obviously, included the local seafood delicacy of greenshell mussels.[10] Except for a curious few who lingered to read the panels, the crowd dispersed and Havelock township went back to sleep.

Its moment in the glare of public attention had passed but now, where for many years a shipwreck memorial had stood largely ignored and neglected, a beautiful new edifice testified to a remarkable coincidence that linked this remote spot in faraway New Zealand to seminal events in the world of modern science.[11] The event was widely reported in New Zealand in both press and television news programs.[12]

As a complementary part of the celebrations, the University of Canterbury had planned a big public function for its illustrious former student. Although he had spent only one year at the university prior to moving to California, he had developed close ties with the faculty in recent years. Now it planned to

Dr. William H. Pickering, O.N.Z., K.B.E., was made an Honorary Member of the Order of New Zealand. Los Angeles, California, September 2003 (Photo: Pickering Family Collection).

The great public interest engendered by Pickering's visit and his subsequent Order of New Zealand award, prompted an enterprising young New Zealand film producer to approach him toward the end of the year with a request for an on-camera series of interviews on location in and around JPL. It was to be part of a documentary on New Zealand's relationship to the Space Age. Pickering readily agreed and the filming took place at JPL a few months later.[16]

Around the same time, this book project began to capture Pickering's interest as the manuscript took shape. The lengthy, in-depth interviews had been completed and by March 2004 when the author began delivering the opening chapters for his approval, Pickering took great pleasure in the detailed reconstruction of his early life and in sharing the material with Inez. Ever the old egotist, he was clearly looking forward to seeing his name on the shiny jacket of a new book—"perhaps a bestseller, perhaps a movie someday," he joked.

And then, quite suddenly, he was gone.

William Pickering died of pneumonia at his home in La Cañada-Flintridge, California, on 15 March 2004, by a remarkable coincidence one year to the day after the dedication ceremony for the memorial in Havelock, New Zealand.

His untimely death sent shockwaves around the world and was reported widely at home and abroad. Wherever his accomplishments in space had been recognized throughout the last half of the 20th century, dozens of newspapers, journals, and magazines published accolades to his brilliant career.

More than any Other . . .

A few days later, family, friends, and colleagues from many walks of life gathered in the beautiful Beckman Auditorium on the Caltech campus in Pasadena for a quiet memorial service to pay a final tribute to William Pickering. There, amid the palm and olive trees, alcoves and columns of the stuccoed Spanish-style buildings that he loved so well, he would have felt completely at home. New Zealand Consul General Darryl Dunn, JPL Director Charles Elachi, and former colleagues joined Caltech President Emeritus Thomas Everhart in eulogizing William Hayward Pickering as they remembered him. All perceived the character of the man in their individual ways; all praised his gentleness, avid determination to succeed, his concern for humanity, and, above all, they praised him for his integrity and adherence to the fundamental principles of scientific enquiry. "In his personal life as in his professional life in the world of space science and technology," they said, "William Pickering had set standards of excellence that would be an example for all that would surely follow."

On an earlier occasion recognizing William Pickering's achievements, President Emeritus Thomas Everhart had written, "More than any other individual, Bill Pickering was responsible for America's success in exploring the planets—an endeavor that demanded vision, courage, dedication, expertise, and the ability to inspire two generations of scientists and engineers at the Jet Propulsion Laboratory."[17]

That tribute still stood. William Pickering would have liked that—it was epitaph enough.

Endnotes

1 Pickering W. H., Interview with the author, Part 7C, August 2003. In the Pickering Collection, Pasadena Museum of History, Pasadena, California.

2 Pickering, W. H., "Early Solar System Exploration." 53rd International Astronautical Congress, Houston, Texas, October 2002. In the Pickering Collection, Pasadena Museum of History, Pasadena, California.

3 Pickering, W. H., "Opening remarks to the International Telemetering Conference." International Telemetering Conference, Las Vegas, Nevada, October 2003. In the Pickering Collection, Pasadena Museum of History, Pasadena, California.

4 His subject for the public town hall lecture was "Challenges of Exploring the Solar System with Automated Spacecraft." In the Pickering Collection, Pasadena Museum of History, Pasadena, California.

5 "Gifford Observatory Opened by Inter-Stellar Scientist." *The Lampstand*, Wellington College, September 2002. In the Pickering Collection, Pasadena Museum of History, Pasadena, California.

6 His subject for the IPENZ lecture was "Managing Innovation—Reflections on a Career." In the Pickering Collection, Pasadena Museum of History, Pasadena, California.

7 Newspaper reports of his visit to Wellington were carried in *The Evening Post*, 31 January 2002; The Dominion, 22 March 2002; and *The Evening Post*, 21 March 2002.

8 Some distance away, the old Havelock Primary Schoolhouse that Pickering attended now served as a youth hostel for young people visiting the area.

9 Pickering, W. H., "Havelock, New Zealand," 15 March 2003; Private correspondence with the author, March 2003. In the Pickering Collection, Pasadena Museum of History, Pasadena, California.

10 McIntyre, Anna, "Old friends reunited at Havelock function." *The Marlborough Express*, 17 March 2003.

11 Campbell, John, "Exploring the Planets–NZ's Space Pioneer." *The Marlborough Express*, 14 March 2003.

12 University of Canterbury, "Canterbury pays tribute to famous space engineer alumnus." Chronicle, vol. 38, no.4, March 2003.

13 He was alluding to the recently established Pickering fellowship that provided support for that purpose.

14 Henzell, John, "Space man returns to Canterbury." *The Press*, Christchurch, New Zealand, 19 March 2003.

15 NZPA. "Queen's Birthday Honours." New Zealand Herald, 9 June 2003.

16 The documentary film by Gillian Ashurst titled "Rocketman" was eventually shown on New Zealand television TVNZ in September 2004. In the Pickering Collection, Pasadena Museum of History, Pasadena, California.

17 Everhart, Thomas E., "Notes for the François-Xavier Bagnoud Aerospace Prize." Ann Arbor Michigan, October 1993. In the Pickering Collection, Pasadena Museum of History, Pasadena, California.

Chapter 11

In Retrospect

It is timely now to consider how history should regard William Hayward Pickering and to consider by what criteria and what yardstick his contribution to society should be measured. Should he be judged by the standards of his times or the standards of the present, by his contribution to society, or to science or human knowledge, by his contribution to his native country or his adopted country or the world in general? And what is his legacy to the age of space in whose gestation and birth he played such a prominent part?

William Pickering first came to the attention of the world in January 1958 when the media triumphantly announced the successful launch of Explorer 1, the American response to the Soviet deployment a few months earlier of the first Earth-orbiting satellite Sputnik. Along with Wernher von Braun and James Van Allen, William Pickering shared the limelight and the accolades. In that instant of time the Space Age was born and with it the professional reputation of William H. Pickering.

Under Pickering's leadership, JPL designed, built, and dispatched NASA's first Ranger spacecraft to take close-up pictures of the surface of the Moon. Building on its Ranger experience, JPL sent the first spacecraft to Venus and, as technology improved, to Mars. The scientific data returns from each successive mission greatly increased our understanding of the composition and dynamics of the solar system and its planets. When he retired as Director in 1976, Pickering had presided over NASA-JPL's missions to the Moon, Venus, and Mars and laid the basis for the fabulous Voyager Grand Tour of all the planets that would sound the praises of NASA-JPL for the next 25 years. Not all of the missions were successful, but Pickering accepted the responsibility that devolved from his position as Director, regardless of the outcome.

Why did the achievements for which the media gave him credit make him so outstanding and so unique? By current standards, when satellites and planetary missions are relatively commonplace, he would be judged as "one among many." But for those few years of 1957 to 1977, missions to the planets were not commonplace; they were unique in the extreme and drew an enormous amount of worldwide public interest. Judged by the standards of that time in the field of space he was indeed "one among few."

While Pickering would be quick to acknowledge that he was merely a part of the NASA-JPL team that conceived and carried out these exquisite missions to the Moon and planets, the public preferred a hero before a team and Pickering had the restrained charisma of a public hero. Rightly or wrongly, he became the chosen one to be credited with leading the U.S. struggle with the Soviets for eminence in space, specifically in the field of planetary exploration.

This perception of Pickering as a pioneer in deep space was enhanced by his extensive agenda of public speaking. Pickering delivered carefully crafted presentations, aided in many cases by a very competent speechwriter, to give him a gravitas that extended well beyond the field of advanced aerospace engineering that made such missions possible.

Among the international community of nations interested in the field of deep space exploration, he came to represent a standard of excellence and a level of achievement that, emboldened by the awesome resources of NASA and enhanced by the enviable reputation of Caltech, none could emulate. For his native country he became an icon, a role model of the country lad whose basic New Zealand education led him to the esoteric field of cosmic ray physics, and, eventually, to a leading part in opening the new age of space.

Many of his colleagues perceived an obverse side to Pickering's attractive public persona that was abrasive, stubborn, and self-centered. Historians observed that the hubris that JPL staff, Pickering included, frequently exhibited toward their counterparts at NASA Headquarters were well-founded.

But Pickering was very loyal to his senior staff that, in reality, formed the main engine that drove the Laboratory. They were the real source of power and innovation at JPL. Pickering knew that and he realized that they resented interference, direction, and judgment from their civil service administrators at NASA who some perceived as less than worthy of the position of authority they claimed. To keep his people focused and productive under the JPL umbrella, Pickering believed that he had to "protect" them from NASA, and to preserve the freedoms associated with the university campus-like environment that attracted them to JPL in the first place. In doing so, he invoked the wrath of NASA and came very close to terminating his career at JPL.

It was not all one-sided, however, and the NASA-JPL working relationship, abrasive though it was, did little to impair the successful progress of NASA's planetary programs. As a former NASA Associate Administrator remarked, "It took strong efforts by men of good will on both sides to make it work."

Apart from several papers dealing with his cosmic ray research at Caltech, Pickering published relatively few scientific papers. Most of what he did at

JPL prior to its becoming part of NASA was classified and not available for public dissemination. However, as Director of a new NASA facility, he used the prestige of his position to give public expression to his ideas, opinions, and experience and to further the public and government interest in space.

To this end, he delivered a great many public lectures where his natural aptitude for public speaking, dramatic subject matter (space exploration), sense of humor, down-to-earth demeanor, and unique New Zealand accent endeared him to the media and charmed audiences wherever he went. By contrast, on a person-to-person level he was rather intimidating. The depth of his technical and scientific knowledge, complemented by his extensive practical experience, penetrating and logical thought processes, sparse conversation, and direct manner made him a formidable manager for his senior staff and a powerful adversary for his NASA detractors.

Writing on the career of William Pickering in 1965, a leading New York newspaper described his efforts to encourage public and government support for the nation's space program and his publicly expressed confidence in the nation's ability to overcome the apparent Russian dominance in space occasioned by the 1957 Sputnik affair as his "greatest contribution." A similar thought was reflected in the (1993) remarks of a former Caltech president, "More than any other individual, Bill Pickering was responsible for America's success in exploring the planets."

For the U.S. space program, his legacy is exemplified by these two opinions written independently 28 years apart, one at the beginning and the other at the end of his career. It might be argued that others could have led the U.S. into deep space with equal success. Perhaps, but the fact remains that they did not and he did. In a word, he was the right man at the right time.

For humankind's first venture beyond the constraints of our home planet, he led the team that led the way. "To be first" was ever his credo.

Bibliography

American Institute of Aeronautics and Astronautics. AIAA at 50 (New York: AIAA, 1981).

Beasley, A. W. *The Light Accepted: 125 Years of Wellington College* (New Zealand: Wellington College, 1992).

Bragg, J. W. "Development of the Corporal: The Embryo of the Army Missile Program." Army Ballistic Missile Agency, Redstone Arsenal, Historical Monograph No. 4, April 1961.

Bille, Matt and Erika Lishock. *The First Space Race* (College Station, Texas: A&M University Press, 2004).

Bunn, Stuart. *Hayward Heritage: The Story of a Pioneer Family* (Mosgiel, New Zealand: Stuart Bunn, 1996).

Burrows, William E. *This New Ocean* (New York: Random House, 1998).

Campbell, John. *Rutherford: Scientist Supreme* (Christchurch, New Zealand: American Astronautical Society Publications, 1999).

Congdon, Eldred. *A Century of Education in Havelock* (Blenheim, New Zealand: Marlborough Historical Society, 1961).

Dougherty, Ian. *Ham Shacks, Brass Pounders and Rag Chewers: A History of Amateur Radio in New Zealand* (New Zealand: New Zealand Association of Radio Transmitters, 1997).

Dunn, Louis. "The Sergeant Surface to Surface Guided Missile System." Pasadena, California: Jet Propulsion Laboratory, Progress Report No. 20-76, April 1954.

Emme, Eugene M. *Aeronautics and Astronautics: An American Chronology of Science and Technology in the Exploration of Space: 1915-1960* (Washington, DC: National Aeronautics and Space Administration, 1961).

Ezell, Edward C., and Linda Neumann Ezell. *On Mars: Exploration of the Red Planet, 1958-1978* (Washington, DC: NASA SP-4212, 1984).

Gatland, Kenneth. *Space Technology: a Comprehensive History of Space Exploration* (London: Salamander Books Ltd. 1981).

Goodstein, Judith R. *Millikan's School: A History of the California Institute of Technology* (New York: Norton, 1991).

Gorn, Michael H. *The Universal Man: Theodore von Kármán's Life in Aeronautics* (Washington and London: Smithsonian Institution Press, 1992).

Hall, R. Cargill. *Lunar Impact: A History of Project Ranger* (Washington, DC: NASA SP-4210, 1977).

Harford, James. *Korolev: How One Man Masterminded the Soviet Drive to Beat America to the Moon* (New York: John Wiley and Sons, 1997).

Hibbs, Albert R., ed. "Exploration of the Moon, Planets, and Interplanetary Space." Pasadena, California: JPL Technical Report 30-1, April 1959.

Horowitz, Norman H. *To Utopia and Back: The Search for Life in the Solar System* (New York: W. H. Freeman and Company, 1986).

Jet Propulsion Laboratory. "Abstracts of Papers Presented at the Guided Missiles and Upper Atmosphere Symposium, 13–16 March 1946." Pasadena, California: JPL Report No.15, April 1946 (unpublished).

Kargon, Robert. *The Rise of Robert Millikan: Portrait of a Life in American Science* (Ithaca: Cornell University Press, 1982).

Kármán, Theodore von, with Lee Edson. *The Wind and Beyond: Theodore von Kármán, Pioneer in Aviation and Pathfinder in Space* (Boston: Little, Brown and Company, 1967).

Kluger, Jeffrey. *Journey Beyond Selene; Remarkable Expeditions Past Our Moon and to the Ends of the Solar System* (New York: Simon and Schuster, 1999).

Koppes, Clayton R. *JPL and the American Space Program: A History of the Jet Propulsion Laboratory* (New Haven: Yale University Press, 1982).

Logsdon, John M., ed. *Exploring the Unknown: Selected Documents in the History of the U.S. Civil Space Program* (Washington, DC: NASA History Series, NASA SP-4407, 2001).

Lord, M.G. *Astro Turf: The Private Life of Rocket Science* (New York: Walker and Company, 2005).

Macomber, H. L., and Wilson, J. H. "California Four Cities Program; 1971-73." Pasadena, California: Jet Propulsion Laboratory, 15 May 1974.

Miles, Ralf F., ed. *Systems Concepts: Lectures on Contemporary Approaches to Systems* (New York: John Wiley & Sons, 1973).

Mudgway, Douglas J. *Uplink-Downlink: A History of the Deep Space Network* (Washington, DC: NASA SP-4227, 2001).

Mudgway, Douglas J. *Big Dish: Building America's Deep Space Connection to the Planets* (Gainesville: University Press of Florida, 2005).

Murray, Bruce. *Journey Into Space: The First Thirty Years of Space Exploration* (New York: W. W. Norton, 1989).

Newell, Homer E. *Beyond the Atmosphere: Early Years of Space Science* (Washington, DC: NASA SP-4211, 1980).

Pickering, W. H. "Collected letters of William Pickering to his Father: 1923-1927." Beth Pickering Mezitt in correspondence with the author.

Pickering, W. H. "Control and Telemetering for Corporal-E." Pasadena, California: Jet Propulsion Laboratory Progress Report No. 4-15, May 1945.

Pickering, W. H., and J. A. Young. "External Instrumentation for NAMT." Pasadena, California: Jet Propulsion Laboratory Progress Report No. 10-1 to 10-9, 1946–1947.

Pickering, W. H. "Study of the Upper Atmosphere by Means of Rockets." Pasadena, California: Jet Propulsion Laboratory, Publication No. 15, June 1947.

Pickering, W. H., and P. H. Reedy. "External Instrumentation for NAMTC." Pasadena, California: Jet Propulsion Laboratory, Progress Report No. 18-1 to 18-8, 1949–1950.

Pickering, W. H. "The Corporal: A Surface-to-Surface Guided Ballistic Missile." Pasadena, California: Jet Propulsion Laboratory, Progress Report No. 20-100, March 1958.

Pickering, William H., and Robert J. Parks. "Guidance and Control System." Alexandria, Virginia: United States Patent Office, Number 3,179,353. 20 April 1965.

Pickering, William H., with James H. Wilson. "Countdown to Space Exploration: A Memoir of the Jet Propulsion Laboratory, 1944-1958" in Hall, ed., *History of Rocketry and Astronautics* (Springfield, Virginia: American Astronautical Society History Series, Vol. 7, Part II, October 1972).

Pickering Research Corporation. "Space and Missile Reliability and Safety Programs." Nuclear Safety Analysis Center, Doc. NSAC 31, February 1981.

Redstone Arsenal. "Historical Monographs on Corporal and Sergeant." On-line versions in PDF format are found online at *http://www.redstone.army.mil/history/pdf/welcome.html* (accessed 11 June 2007).

Research Institute. "... And Thus Seek Knowledge" (Dhahran, Saudi Arabia: University of Petroleum and Minerals, January 1976).

Research Institute. "The Ultimate Resource" (Dhahran, Saudi Arabia: University of Petroleum and Minerals, November 1979).

Scheid, Ann. *Pasadena: Crown of the Valley* (Northridge: Windsor Publications, 1986).

Schorn, Ronald A. *Planetary Astronomy; From Ancient Times to the Third Millennium* (College Station, Texas: A&M University Press, 1998).

Siddiqi, Asif A. *Deep Space Chronicle: A Chronology of Deep Space and Planetary Probes, 1958-2000* (Washington, DC: NASA SP-4524, 2002).

Swanson, Glen E. *Before This Decade Is Out: Personal Reflections on the Apollo Program* (Washington, DC: NASA SP-4233, 1999).

Thomas, Shirley. *Men in Space, Vol. 2* (Philadelphia: Chilton Company, 1962).

Time Magazine, "Voyage to the Morning Star." Vol. 81, No. 10 (8 March 1963).

Time Magazine, "Portrait of a Planet." Vol. 86, No. 4 (23 July 1965).

Van Allen, James A. "Radiation Belts Around the Earth." *Scientific American*, 200, March 1959.

Wellington College Archives. "The Wellingtonian." Wellington, New Zealand: Vol. 33, December 1924; Vol. 34, December 1925; Vol. 35, December 1926; Vol. 36, December 1927.

Weaver, Kenneth F. "Journey to Mars." *National Geographic*, February 1973.

Westwick, Peter J. *Into the Black: JPL and the American Space Program, 1976–2004* (New Haven: Yale University Press, 2007).

Wheelock, Harold, J., ed. *Mariner: Mission to Venus* (New York: McGraw Hill, 1963).

Wilson, James H. Interviews with William H. Pickering. Pasadena, California: Jet Propulsion Laboratory, Oral History Program, 1972.

A Note on Sources

Much of the background material for this account of the life of William H. Pickering is based on Pickering's interviews with the author extending over the 15-month period that preceded his death in March 2004. The tapes and transcriptions of these interviews are referred to in the text as "Interviews with Author" and are included with Pickering's personal papers held by the Pasadena Museum of History in its William H. Pickering Collection. The personal papers span Pickering's life from childhood in New Zealand through the end of his business life in the U.S.

Additional material on the New Zealand aspects of Pickering's life may be found in the Archives at Wellington College, and in the Pickering Collection at the Alexander Turnbull Library in Wellington, New Zealand.

The extensive JPL Archival holdings on William H. Pickering cover his professional life as director of JPL, from 1958 through his retirement in 1976. They are arranged in several numbered, processed collections, as follows:

Pickering (William H.) Collection, 1958-1976, JPL214

Pickering (William H.) Committee Organizations Collection, 1962–1970, JPL140

Pickering (William H.) Office File Collection, 1955–1976, JPL186

Pickering (William H.) Publications Collection, 1932–1971, JPL133

Pickering (William H.) Records, February 1970 to March 1989, JPL3

Pickering (William H.) Speech Collection 1955–1975, JPL181

Pickering (William H.) Speech Reference Collection, 1959–1974, JPL187

Available online at *http://beacon.jpl.nasa.gov/Find/FindHistorical/archlist.htm*

Additional archival material relating to William H. Pickering can be found in the JPL Archives History Collection online at *http://beacon.jpl.nasa.gov/Find/FindArchivesCat.htm*

The Archives of the California Institute of Technology also contain an extensive processed collection of the papers of William H. Pickering spanning the period 1941 to 1970. Titled "The Papers of William H. Pickering (1910-present)," the collection was processed by Laurence M. Dupray and Joy A. Pinter in June 2001. The collection is arranged serially as follows:

Series I General Correspondence

Series II Jet Propulsion Laboratory (includes Caltech and NASA-related items)

Series III Conferences

Series IV Professional Organizations

Series V Committees

Series VI Awards and Miscellaneous

Available online at *http://archives.caltech.edu/collections.cfm*

The institutional business of JPL and its interaction with NASA and Caltech under the direction of William H. Pickering has been well documented by Clayton R. Koppes in his detailed history of JPL, "JPL and the American Space Program," Yale University Press, 1982. Where necessary, the narrative draws on Koppes' distillation of archival material rather than the primary sources to illustrate significant events in the history of the laboratory during the Pickering years.

Index

A

Advanced Research Projects Agency (ARPA), 86, 91

Age of Space, arrives, 88

Air Corps, 42-44

Altadena, California, 49, 59, 67, 114-115, 220

American Astronautical Society (AAS), 103, 173

American Institute of Aeronautics and Astronautics (AIAA), 108, 124, 162, 207

American Rocket Society, (ARS), 63, 100, 107-108, 124

Ames Research Center (ARC), 162

Apollo Program, 109, 155, 163

Army Ballistic Missile Agency (ABMA), 62, 76, 91

Arnold, H. H. "Hap", General, 42, 48-49

Arthur, George, 103

Arroyo Seco, 42-45, 213, 220

Arrival, in California, 22

Association of Engineering Societies, 123

Aviation Week, 103

Awards, citations, distinctions, etc., 123, 138, 147-148, 155, 159, 173, 186, 205-208, 210, 220-221, 224-226

B

Bagnoud, François-Xavier Aerospace Prize, 208-209

Beeman, Arthur, 190

Barraclough, Mr., 6

Berkner, Lloyd, 74, 79

Blagonravov, Anatoly; report to CSAGI, 74

Bouchey, Lt. Homer, 43

British Interplanetary Society, 148, 173

Brown, Harold, 171, 189, 191-192

Burke, James, D., 114

C

California Institute of Technology; academic achievements at, 23, 26, 33; and JPL, 44, 45, 50, 160, 171; and NASA, 125, 160-161; origin of, 21, 33; professor of electrical engineering at, 33-34, 49, 50, 59, 189; and William Pickering, 22, 160, 171

Caltech, (See California Institute of Technology)

Campbell, John, 2, 3, 222, 224

Canterbury, University of, 225; academic achievements at, 16; origin of, 16; confers honorary doctorate, 224

Christchurch, New Zealand, 2, 14-15, 146, 224-225

Civil Systems, Caltech attitude to, 171; projects at JPL (See Technology Transfer), 167, 174; impact on JPL, 173, 177-178, 188

Clark, John F.; as general manager at JPL, 158

Cold War Warrior, 41

Columbus Medal, 138

Committee for Space Research (COSPAR), 186

Comité Speciale de l'Année Géophysique Internationale (CSAGI), 73

Coincidence Detectors, (See Cosmic Ray Researcher)

Congressional Reviews; for Ranger 6, 129; for Surveyor, 156; and JPL visits, 102

Corporal program; IRBM prototype, 55, 56; testing at White Sands, 57-60; "Round 100", 60; deployment in Europe, 58

Corporal research test missile, 45, 47, 50-51, 55

Cosmic Ray Researcher, 21, 26; cosmic ray telescope, 31; Geiger counter tubes, 26-28; Coincidence detectors, 27-28, 30; telemetry for high altitude balloons, 34; improvements in experimental techniques, 26, 27-29; thesis for Ph.D., 30-31; search for latitude effect, 31-32; limitations of rockets, 62

Cummings, Clifford I., 50, 114

D

Dabbagh, Abdullah, 199-200

Deep Space Network, 117, 150; founding of, 91, 92; largest antennas of, 165-166, 181, 184

Departure for California, 17

Department of Defense, 54, 76

Documentary film produced, 226

Douslin, Horace, 10, 13-14, 16-17

Douslin, Kate, 4

Drag Brakes; demonstrating the principle of, 57

Index

NASA History Series

Reference Works, NASA SP-4000:

Grimwood, James M. *Project Mercury: A Chronology*. NASA SP-4001, 1963.

Grimwood, James M., and Barton C. Hacker, with Peter J. Vorzimmer. *Project Gemini Technology and Operations: A Chronology*. NASA SP-4002, 1969.

Link, Mae Mills. *Space Medicine in Project Mercury*. NASA SP-4003, 1965.

Astronautics and Aeronautics, 1963: Chronology of Science, Technology, and Policy. NASA SP-4004, 1964.

Astronautics and Aeronautics, 1964: Chronology of Science, Technology, and Policy. NASA SP-4005, 1965.

Astronautics and Aeronautics, 1965: Chronology of Science, Technology, and Policy. NASA SP-4006, 1966.

Astronautics and Aeronautics, 1966: Chronology of Science, Technology, and Policy. NASA SP-4007, 1967.

Astronautics and Aeronautics, 1967: Chronology of Science, Technology, and Policy. NASA SP-4008, 1968.

Ertel, Ivan D., and Mary Louise Morse. *The Apollo Spacecraft: A Chronology, Volume I, Through November 7, 1962*. NASA SP-4009, 1969.

Morse, Mary Louise, and Jean Kernahan Bays. *The Apollo Spacecraft: A Chronology, Volume II, November 8, 1962–September 30, 1964*. NASA SP-4009, 1973.

Brooks, Courtney G., and Ivan D. Ertel. *The Apollo Spacecraft: A Chronology, Volume III, October 1, 1964–January 20, 1966*. NASA SP-4009, 1973.

Ertel, Ivan D., and Roland W. Newkirk, with Courtney G. Brooks. *The Apollo Spacecraft: A Chronology, Volume IV, January 21, 1966–July 13, 1974*. NASA SP-4009, 1978.

Astronautics and Aeronautics, 1968: Chronology of Science, Technology, and Policy. NASA SP-4010, 1969.

Newkirk, Roland W., and Ivan D. Ertel, with Courtney G. Brooks. *Skylab: A Chronology*. NASA SP-4011, 1977.

Van Nimmen, Jane, and Leonard C. Bruno, with Robert L. Rosholt. *NASA Historical Data Book, Vol. I: NASA Resources, 1958–1968*. NASA SP-4012, 1976, rep. ed. 1988.

Ezell, Linda Neuman. *NASA Historical Data Book, Vol. II: Programs and Projects, 1958–1968*. NASA SP-4012, 1988.

Ezell, Linda Neuman. *NASA Historical Data Book, Vol. III: Programs and Projects, 1969–1978*. NASA SP-4012, 1988.

Gawdiak, Ihor, with Helen Fedor. *NASA Historical Data Book, Vol. IV: NASA Resources, 1969–1978*. NASA SP-4012, 1994.

Rumerman, Judy A. *NASA Historical Data Book, Vol. V: NASA Launch Systems, Space Transportation, Human Spaceflight, and Space Science, 1979–1988*. NASA SP-4012, 1999.

Rumerman, Judy A. *NASA Historical Data Book, Vol. VI: NASA Space Applications, Aeronautics and Space Research and Technology, Tracking and Data Acquisition/Support Operations, Commercial Programs, and Resources, 1979–1988*. NASA SP-4012, 1999.

Astronautics and Aeronautics, 1969: Chronology of Science, Technology, and Policy. NASA SP-4014, 1970.

Astronautics and Aeronautics, 1970: Chronology of Science, Technology, and Policy. NASA SP-4015, 1972.

Astronautics and Aeronautics, 1971: Chronology of Science, Technology, and Policy. NASA SP-4016, 1972.

Astronautics and Aeronautics, 1972: Chronology of Science, Technology, and Policy. NASA SP-4017, 1974.

Astronautics and Aeronautics, 1973: Chronology of Science, Technology, and Policy. NASA SP-4018, 1975.

Astronautics and Aeronautics, 1974: Chronology of Science, Technology, and Policy. NASA SP-4019, 1977.

Astronautics and Aeronautics, 1975: Chronology of Science, Technology, and Policy. NASA SP-4020, 1979.

Astronautics and Aeronautics, 1976: Chronology of Science, Technology, and Policy. NASA SP-4021, 1984.

Astronautics and Aeronautics, 1977: Chronology of Science, Technology, and Policy. NASA SP-4022, 1986.

Astronautics and Aeronautics, 1978: Chronology of Science, Technology, and Policy. NASA SP-4023, 1986.

Astronautics and Aeronautics, 1979–1984: Chronology of Science, Technology, and Policy. NASA SP-4024, 1988.

Astronautics and Aeronautics, 1985: Chronology of Science, Technology, and Policy. NASA SP-4025, 1990.

Noordung, Hermann. *The Problem of Space Travel: The Rocket Motor.* Edited by Ernst Stuhlinger and J.D. Hunley, with Jennifer Garland. NASA SP-4026, 1995.

Astronautics and Aeronautics, 1986–1990: A Chronology. NASA SP-4027, 1997.

Astronautics and Aeronautics, 1991–1995: A Chronology. NASA SP-2000-4028, 2000.

Orloff, Richard W. *Apollo by the Numbers: A Statistical Reference.* NASA SP-2000-4029, 2000.

Management Histories, NASA SP-4100:

Rosholt, Robert L. *An Administrative History of NASA, 1958–1963.* NASA SP-4101, 1966.

Levine, Arnold S. *Managing NASA in the Apollo Era.* NASA SP-4102, 1982.

Roland, Alex. *Model Research: The National Advisory Committee for Aeronautics, 1915–1958.* NASA SP-4103, 1985.

Fries, Sylvia D. *NASA Engineers and the Age of Apollo.* NASA SP-4104, 1992.

Glennan, T. Keith. *The Birth of NASA: The Diary of T. Keith Glennan.* Edited by J.D. Hunley. NASA SP-4105, 1993.

Seamans, Robert C. *Aiming at Targets: The Autobiography of Robert C. Seamans.* NASA SP-4106, 1996.

Garber, Stephen J., editor. *Looking Backward, Looking Forward: Forty Years of Human Spaceflight Symposium.* NASA SP-2002-4107, 2002.

Mallick, Donald L. with Peter W. Merlin. *The Smell of Kerosene: A Test Pilot's Odyssey.* NASA SP-4108, 2003.

Iliff, Kenneth W. and Curtis L. Peebles. *From Runway to Orbit: Reflections of a NASA Engineer.* NASA SP-2004-4109, 2004.

Chertok, Boris. *Rockets and People, Volume 1.* NASA SP-2005-4110, 2005.

Chertok, Boris. *Rockets and People: Creating a Rocket Industry, Volume II*. NASA SP-2006-4110, 2006.

Laufer, Alexander, Todd Post, and Edward Hoffman. *Shared Voyage: Learning and Unlearning from Remarkable Projects*. NASA SP-2005-4111, 2005.

Dawson, Virginia P., and Mark D. Bowles. *Realizing the Dream of Flight: Biographical Essays in Honor of the Centennial of Flight, 1903–2003*. NASA SP-2005-4112, 2005.

Mudgway, Douglas J. *William H. Pickering: America's Deep Space Pioneer*. NASA SP-2007-4113,

Project Histories, NASA SP-4200:

Swenson, Loyd S., Jr., James M. Grimwood, and Charles C. Alexander. *This New Ocean: A History of Project Mercury*. NASA SP-4201, 1966; reprinted 1999.

Green, Constance McLaughlin, and Milton Lomask. *Vanguard: A History*. NASA SP-4202, 1970; rep. ed. Smithsonian Institution Press, 1971.

Hacker, Barton C., and James M. Grimwood. *On Shoulders of Titans: A History of Project Gemini*. NASA SP-4203, 1977, reprinted 2002.

Benson, Charles D., and William Barnaby Faherty. *Moonport: A History of Apollo Launch Facilities and Operations*. NASA SP-4204, 1978.

Brooks, Courtney G., James M. Grimwood, and Loyd S. Swenson, Jr. *Chariots for Apollo: A History of Manned Lunar Spacecraft*. NASA SP-4205, 1979.

Bilstein, Roger E. *Stages to Saturn: A Technological History of the Apollo/Saturn Launch Vehicles*. NASA SP-4206, 1980 and 1996.

Compton, W. David, and Charles D. Benson. *Living and Working in Space: A History of Skylab*. NASA SP-4208, 1983.

Ezell, Edward Clinton, and Linda Neuman Ezell. *The Partnership: A History of the Apollo-Soyuz Test Project*. NASA SP-4209, 1978.

Hall, R. Cargill. *Lunar Impact: A History of Project Ranger*. NASA SP-4210, 1977.

Newell, Homer E. *Beyond the Atmosphere: Early Years of Space Science*. NASA SP-4211, 1980.

Ezell, Edward Clinton, and Linda Neuman Ezell. *On Mars: Exploration of the Red Planet, 1958–1978*. NASA SP-4212, 1984.

Pitts, John A. *The Human Factor: Biomedicine in the Manned Space Program to 1980*. NASA SP-4213, 1985.

Compton, W. David. *Where No Man Has Gone Before: A History of Apollo Lunar Exploration Missions*. NASA SP-4214, 1989.

Naugle, John E. *First Among Equals: The Selection of NASA Space Science Experiments*. NASA SP-4215, 1991.

Wallace, Lane E. *Airborne Trailblazer: Two Decades with NASA Langley's 737 Flying Laboratory*. NASA SP-4216, 1994.

Butrica, Andrew J., ed. *Beyond the Ionosphere: Fifty Years of Satellite Communications*. NASA SP-4217, 1997.

Butrica, Andrew J. *To See the Unseen: A History of Planetary Radar Astronomy.* NASA SP-4218, 1996.

Mack, Pamela E., ed. *From Engineering Science to Big Science: The NACA and NASA Collier Trophy Research Project Winners.* NASA SP-4219, 1998.

Reed, R. Dale. *Wingless Flight: The Lifting Body Story.* NASA SP-4220, 1998.

Heppenheimer, T. A. *The Space Shuttle Decision: NASA's Search for a Reusable Space Vehicle.* NASA SP-4221, 1999.

Hunley, J. D., ed. *Toward Mach 2: The Douglas D-558 Program.* NASA SP-4222, 1999.

Swanson, Glen E., ed. *"Before This Decade is Out . . ." Personal Reflections on the Apollo Program.* NASA SP-4223, 1999.

Tomayko, James E. *Computers Take Flight: A History of NASA's Pioneering Digital Fly-By-Wire Project.* NASA SP-4224, 2000.

Morgan, Clay. *Shuttle-Mir: The United States and Russia Share History's Highest Stage.* NASA SP-2001-4225.

Leary, William M. *We Freeze to Please: A History of NASA's Icing Research Tunnel and the Quest for Safety.* NASA SP-2002-4226, 2002.

Mudgway, Douglas J. *Uplink-Downlink: A History of the Deep Space Network, 1957–1997.* NASA SP-2001-4227.

Dawson, Virginia P., and Mark D. Bowles. *Taming Liquid Hydrogen: The Centaur Upper Stage Rocket, 1958–2002.* NASA SP-2004-4230.

Meltzer, Michael. *Mission to Jupiter: A History of the Galileo Project.* NASA SP-2007-4231.

Heppenheimer, T. A. *Facing the Heat Barrier: A History of Hypersonics.* NASA SP-2007-4232.

Tsiao, Sunny. *"Read You Loud and Clear!" The Story of NASA's Spaceflight Tracking and Data Network.* NASA SP-2007-4233.

Center Histories, NASA SP-4300:

Rosenthal, Alfred. *Venture into Space: Early Years of Goddard Space Flight Center.* NASA SP-4301, 1985.

Hartman, Edwin, P. *Adventures in Research: A History of Ames Research Center, 1940–1965.* NASA SP-4302, 1970.

Hallion, Richard P. *On the Frontier: Flight Research at Dryden, 1946–1981.* NASA SP-4303, 1984.

Muenger, Elizabeth A. *Searching the Horizon: A History of Ames Research Center, 1940–1976.* NASA SP-4304, 1985.

Hansen, James R. *Engineer in Charge: A History of the Langley Aeronautical Laboratory, 1917–1958.* NASA SP-4305, 1987.

Dawson, Virginia P. *Engines and Innovation: Lewis Laboratory and American Propulsion Technology.* NASA SP-4306, 1991.

Dethloff, Henry C. *"Suddenly Tomorrow Came . . .": A History of the Johnson Space Center, 1957–1990*. NASA SP-4307, 1993.

Hansen, James R. *Spaceflight Revolution: NASA Langley Research Center from Sputnik to Apollo*. NASA SP-4308, 1995.

Wallace, Lane E. *Flights of Discovery: An Illustrated History of the Dryden Flight Research Center*. NASA SP-4309, 1996.

Herring, Mack R. *Way Station to Space: A History of the John C. Stennis Space Center*. NASA SP-4310, 1997.

Wallace, Harold D., Jr. *Wallops Station and the Creation of an American Space Program*. NASA SP-4311, 1997.

Wallace, Lane E. *Dreams, Hopes, Realities. NASA's Goddard Space Flight Center: The First Forty Years*. NASA SP-4312, 1999.

Dunar, Andrew J., and Stephen P. Waring. *Power to Explore: A History of Marshall Space Flight Center, 1960–1990*. NASA SP-4313, 1999.

Bugos, Glenn E. *Atmosphere of Freedom: Sixty Years at the NASA Ames Research Center*. NASA SP-2000-4314, 2000.

Schultz, James. *Crafting Flight. Aircraft Pioneers and the Contributions of the Men and Women of NASA Langley Research Center*. NASA SP-2003-4316, 2003.

Bowles, Mark D. *Science in Flux: NASA's Nuclear Program at Plum Brook Station, 1955–2005*. NASA SP-2006-4317.

Wallace, Lane E. *Flights of Discovery: An Illustrated History of the Dryden Flight Research Center*. NASA SP-4318, 2007. Revised version of SP-4309.

General Histories, NASA SP-4400:

Corliss, William R. *NASA Sounding Rockets, 1958–1968: A Historical Summary*. NASA SP-4401, 1971.

Wells, Helen T., Susan H. Whiteley, and Carrie Karegeannes. *Origins of NASA Names*. NASA SP-4402, 1976.

Anderson, Frank W., Jr. *Orders of Magnitude: A History of NACA and NASA, 1915–1980*. NASA SP-4403, 1981.

Sloop, John L. *Liquid Hydrogen as a Propulsion Fuel, 1945–1959*. NASA SP-4404, 1978.

Roland, Alex. *A Spacefaring People: Perspectives on Early Spaceflight*. NASA SP-4405, 1985.

Bilstein, Roger E. *Orders of Magnitude: A History of the NACA and NASA, 1915–1990*. NASA SP-4406, 1989.

Logsdon, John M., ed., with Linda J. Lear, Jannelle Warren Findley, Ray A. Williamson, and Dwayne A. Day. *Exploring the Unknown: Selected Documents in the History of the U.S. Civil Space Program, Volume I, Organizing for Exploration*. NASA SP-4407, 1995.

Logsdon, John M., ed, with Dwayne A. Day, and Roger D. Launius. *Exploring the Unknown: Selected Documents in the History of the U.S. Civil Space Program, Volume II, External Relationships*. NASA SP-4407, 1996.

Logsdon, John M., ed., with Roger D. Launius, David H. Onkst, and Stephen J. Garber. *Exploring the Unknown: Selected Documents in the History of the U.S. Civil Space Program, Volume III, Using Space*. NASA SP-4407,1998.

Logsdon, John M., ed., with Ray A. Williamson, Roger D. Launius, Russell J. Acker, Stephen J. Garber, and Jonathan L. Friedman. *Exploring the Unknown: Selected Documents in the History of the U.S. Civil Space Program, Volume IV, Accessing Space*. NASA SP-4407, 1999.

Logsdon, John M., ed., with Amy Paige Snyder, Roger D. Launius, Stephen J. Garber, and Regan Anne Newport. *Exploring the Unknown: Selected Documents in the History of the U.S. Civil Space Program, Volume V, Exploring the Cosmos*. NASA SP-4407, 2001.

Logsdon, John M., ed., with Stephen J. Garber, Roger D. Launius, and Ray A. Williamson. *Exploring the Unknown: Selected Documents in the History of the U.S. Civil Space Program, Volume VI: Space and Earth Science*. NASA SP-2004-4407, 2004.

Siddiqi, Asif A., *Challenge to Apollo: The Soviet Union and the Space Race, 1945–1974*. NASA SP-2000-4408, 2000.

Hansen, James R., ed. *The Wind and Beyond: Journey into the History of Aerodynamics in America, Volume 1, The Ascent of the Airplane*. NASA SP-2003-4409, 2003.

Hansen, James R., ed. *The Wind and Beyond: Journey into the History of Aerodynamics in America, Volume 2, Reinventing the Airplane*. NASA SP-2007-4409, 2007.

Hogan, Thor. *Mars Wars: The Rise and Fall of the Space Exploration Initiative*. NASA SP-2007-4410, 2007.

Monographs in Aerospace History (SP-4500 Series):

Launius, Roger D., and Aaron K. Gillette, comps. *Toward a History of the Space Shuttle: An Annotated Bibliography*. Monograph in Aerospace History, No. 1, 1992.

Launius, Roger D., and J. D. Hunley, comps. *An Annotated Bibliography of the Apollo Program*. Monograph in Aerospace History No. 2, 1994.

Launius, Roger D. *Apollo: A Retrospective Analysis*. Monograph in Aerospace History, No. 3, 1994.

Hansen, James R. *Enchanted Rendezvous: John C. Houbolt and the Genesis of the Lunar-Orbit Rendezvous Concept*. Monograph in Aerospace History, No. 4, 1995.

Gorn, Michael H. *Hugh L. Dryden's Career in Aviation and Space*. Monograph in Aerospace History, No. 5, 1996.

Powers, Sheryll Goecke. *Women in Flight Research at NASA Dryden Flight Research Center from 1946 to 1995*. Monograph in Aerospace History, No. 6, 1997.

Portree, David S. F., and Robert C. Trevino. *Walking to Olympus: An EVA Chronology*. Monograph in Aerospace History, No. 7, 1997.

Logsdon, John M., moderator. *Legislative Origins of the National Aeronautics and Space Act of 1958: Proceedings of an Oral History Workshop.* Monograph in Aerospace History, No. 8, 1998.

Rumerman, Judy A., comp. *U.S. Human Spaceflight, A Record of Achievement 1961–1998.* Monograph in Aerospace History, No. 9, 1998.

Portree, David S. F. *NASA's Origins and the Dawn of the Space Age.* Monograph in Aerospace History, No. 10, 1998.

Logsdon, John M. *Together in Orbit: The Origins of International Cooperation in the Space Station.* Monograph in Aerospace History, No. 11, 1998.

Phillips, W. Hewitt. *Journey in Aeronautical Research: A Career at NASA Langley Research Center.* Monograph in Aerospace History, No. 12, 1998.

Braslow, Albert L. *A History of Suction-Type Laminar-Flow Control with Emphasis on Flight Research.* Monograph in Aerospace History, No. 13, 1999.

Logsdon, John M., moderator. *Managing the Moon Program: Lessons Learned From Apollo.* Monograph in Aerospace History, No. 14, 1999.

Perminov, V. G. *The Difficult Road to Mars: A Brief History of Mars Exploration in the Soviet Union.* Monograph in Aerospace History, No. 15, 1999.

Tucker, Tom. *Touchdown: The Development of Propulsion Controlled Aircraft at NASA Dryden.* Monograph in Aerospace History, No. 16, 1999.

Maisel, Martin, Demo J.Giulanetti, and Daniel C. Dugan. *The History of the XV-15 Tilt Rotor Research Aircraft: From Concept to Flight.* Monograph in Aerospace History, No. 17, 2000. NASA SP-2000-4517.

Jenkins, Dennis R. *Hypersonics Before the Shuttle: A Concise History of the X-15 Research Airplane.* Monograph in Aerospace History, No. 18, 2000. NASA SP-2000-4518.

Chambers, Joseph R. *Partners in Freedom: Contributions of the Langley Research Center to U.S. Military Aircraft of the 1990s.* Monograph in Aerospace History, No. 19, 2000. NASA SP-2000-4519.

Waltman, Gene L. *Black Magic and Gremlins: Analog Flight Simulations at NASA's Flight Research Center.* Monograph in Aerospace History, No. 20, 2000. NASA SP-2000-4520.

Portree, David S. F. *Humans to Mars: Fifty Years of Mission Planning, 1950–2000.* Monograph in Aerospace History, No. 21, 2001. NASA SP-2001-4521.

Thompson, Milton O., with J. D. Hunley. *Flight Research: Problems Encountered and What they Should Teach Us.* Monograph in Aerospace History, No. 22, 2001. NASA SP-2001-4522.

Tucker, Tom. *The Eclipse Project.* Monograph in Aerospace History, No. 23, 2001. NASA SP-2001-4523.

Siddiqi, Asif A. *Deep Space Chronicle: A Chronology of Deep Space and Planetary Probes 1958–2000.* Monograph in Aerospace History, No. 24, 2002. NASA SP-2002-4524.

Merlin, Peter W. *Mach 3+: NASA/USAF YF-12 Flight Research, 1969–1979.* Monograph in Aerospace History, No. 25, 2001. NASA SP-2001-4525.

Anderson, Seth B. *Memoirs of an Aeronautical Engineer: Flight Tests at Ames Research Center: 1940–1970.* Monograph in Aerospace History, No. 26, 2002. NASA SP-2002-4526.

Renstrom, Arthur G. *Wilbur and Orville Wright: A Bibliography Commemorating the One-Hundredth Anniversary of the First Powered Flight on December 17, 1903.* Monograph in Aerospace History, No. 27, 2002. NASA SP-2002-4527.

No monograph 28.

Chambers, Joseph R. *Concept to Reality: Contributions of the NASA Langley Research Center to U.S. Civil Aircraft of the 1990s.* Monograph in Aerospace History, No. 29, 2003. SP-2003-4529.

Peebles, Curtis, editor. *The Spoken Word: Recollections of Dryden History, The Early Years.* Monograph in Aerospace History, No. 30, 2003. SP-2003-4530.

Jenkins, Dennis R., Tony Landis, and Jay Miller. *American X-Vehicles: An Inventory- X-1 to X-50.* Monograph in Aerospace History, No. 31, 2003. SP-2003-4531.

Renstrom, Arthur G. *Wilbur and Orville Wright: A Chronology Commemorating the One-Hundredth Anniversary of the First Powered Flight on December 17, 1903.* Monograph in Aerospace History, No. 32, 2003. NASA SP-2003-4532.

Bowles, Mark D., and Robert S. Arrighi. *NASA's Nuclear Frontier: The Plum Brook Research Reactor.* Monograph in Aerospace History, No. 33, 2004. (SP-2004-4533).

Matranga, Gene J., C. Wayne Ottinger, Calvin R. Jarvis, and D. Christian Gelzer. *Unconventional, Contrary, and Ugly: The Lunar Landing Research Vehicle.* Monograph in Aerospace History, No. 35, 2006. NASA SP-2004-4535.

McCurdy, Howard E. *Low Cost Innovation in Spaceflight: The History of the Near Earth Asteroid Rendezvous (NEAR) Mission.* Monograph in Aerospace History, No. 36, 2005. NASA SP-2005-4536.

Seamans, Robert C., Jr. *Project Apollo: The Tough Decisions.* Monograph in Aerospace History, No. 37, 2005. NASA SP-2005-4537.

Lambright, W. Henry. *NASA and the Environment: The Case of Ozone Depletion.* Monograph in Aerospace History, No. 38, 2005. NASA SP-2005-4538.

Chambers, Joseph R. *Innovation in Flight: Research of the NASA Langley Research Center on Revolutionary Advanced Concepts for Aeronautics.* Monograph in Aerospace History, No. 39, 2005. NASA SP-2005-4539.

Phillips, W. Hewitt. *Journey Into Space Research: Continuation of a Career at NASA Langley Research Center.* Monograph in Aerospace History, No. 40, 2005. NASA SP-2005-4540.

Rumerman, Judy A., Chris Gamble, and Gabriel Okolski, compilers. *U.S. Human Spaceflight: A Record of Achievement, 1961–2006.* Monograph in Aerospace History No. 41, 2007. NASA SP-2007-4541.

Dryden Historical Studies

Tomayko, James E., author, and Christian Gelzer, editor. *The Story of Self-Repairing Flight Control Systems*. Dryden Historical Study #1.

Electronic Media (SP-4600 Series)

Remembering Apollo 11: The 30th Anniversary Data Archive CD-ROM. NASA SP-4601, 1999.

Remembering Apollo 11: The 35th Anniversary Data Archive CD-ROM. NASA SP-2004-4601, 2004. This is an update of the 1999 edition.

The Mission Transcript Collection: U.S. Human Spaceflight Missions from Mercury Redstone 3 to Apollo 17. SP-2000-4602, 2001. Now available commerically from CG Publishing.

Shuttle-Mir: the United States and Russia Share History's Highest Stage. NASA SP-2001-4603, 2002. This CD-ROM is available from NASA CORE.

U.S. Centennial of Flight Commission presents Born of Dreams ~ Inspired by Freedom. NASA SP-2004-4604, 2004.

Of Ashes and Atoms: A Documentary on the NASA Plum Brook Reactor Facility. NASA SP-2005-4605.

Taming Liquid Hydrogen: The Centaur Upper Stage Rocket Interactive CD-ROM. NASA SP-2004-4606, 2004.

Fueling Space Exploration: The History of NASA's Rocket Engine Test Facility DVD. NASA SP-2005-4607.

Conference Proceedings (SP-4700 Series)

Dick, Steven J., and Keith Cowing, ed. *Risk and Exploration: Earth, Sea and the Stars*. NASA SP-2005-4701.

Dick, Steven J., and Roger D. Launius. *Critical Issues in the History of Spaceflight*. NASA SP-2006-4702.

Societal Impact (SP-4800 Series)

Dick, Steven J., and Roger D. Launius. *Societal Impact of Spaceflight*. NASA SP-2007-4801.